Make: Basic Arduino Projects

26 Experiments with Microcontrollers and Electronics

Don Wilcher

SEBASTOPOL, CA

Make: Basic Arduino Projects

by Don Wilcher

Printed in the United States of America.

Published by Maker Media, Inc., 1005 Gravenstein Highway North, Sebastopol, CA 95472.

Maker Media books may be purchased for educational, business, or sales promotional use. Online editions are also available for most titles (*http://my.safaribooksonline.com*). For more information, contact O'Reilly Media's corporate/institutional sales department: 800-998-9938 or *corporate@oreilly.com*.

Editor: Patrick Di Justo
Production Editor: Kara Ebrahim
Copyeditor: Charles Roumeliotis
Proofreader: Jasmine Kwityn
Indexer: Ellen Troutman
Cover Designer: Juliann Brown
Interior Designer: David Futato
Illustrator: Rebecca Demarest
Photographers: Frank Teng and Don Wilcher

February 2014:   First Edition

Revision History for the First Edition:

2014-02-05:   First release

2014-03-07:   Second release

See *http://oreilly.com/catalog/errata.csp?isbn=9781449360665* for release details.

ISBN: 978-1-449-36066-5

[LSI]

Contents

Preface

So, you've bought the Ultimate Microcontroller Pack to build some cool and fun Arduino projects. Now all you need are some sample projects to build with it! The *Basic Arduino Projects* book is here to help you! It's got a wealth of cool devices and gadgets to build with your Ultimate Microcontroller Pack. The projects in the book explain the world of electronics using a fun and hands-on approach.

The motivation behind writing this book is based on several conversations with Brian Jepson (Make: Books Senior Editor) and the need for a book that allows people to explore the electronic parts and the Arduino within the Ultimate Microcontroller Pack. The Arduino is a very popular Maker platform that allows you to explore electronics with an interactive approach. As awesome as a box of parts is, it's difficult for people with little electronics experience to begin making things with it. This book solves that problem by letting you learn more about electronics while you make fun projects with the parts in this kit. *Basic Arduino Projects* is a practical guide that illustrates how a bunch of electronic parts, coupled with Arduino, can be transformed into awesome devices and gadgets for education and play.

In addition, being an electrical engineer and educator, I'm very sensitive to delivering good instructional content to my students (adults and teenagers). This book was written to attract young readers to the exciting world of electronics by building cool and creative projects using the Ultimate Microcontroller Pack. This book is also intended for Makers and novices who have heard about the Arduino but never experienced the fun and excitement that comes from building cool electronic gadgets and devices with this open hardware platform.

By building and experimenting with the projects in this book, young readers, Makers, and electronic novices will learn how to:

- Read electronic circuit schematic and block diagrams.
- Assemble electronic circuits using the MakerShield prototyping board.

- Build basic logic circuits using the Arduino as a programmable computer brain.
- Use an LCD display for displaying text and special characters.
- Create simple electronic controllers for LEDs and servo motors.

Last, you will learn how to create gadgets and devices for education and play using imagination and the parts from the Ultimate Microcontroller Pack. Enjoy the Maker adventure!

Conventions Used in This Book

The following typographical conventions are used in this book:

Italic
    Indicates new terms, URLs, email addresses, filenames, and file extensions.

`Constant width`
    Used for program listings, as well as within paragraphs to refer to program elements such as variable or function names, databases, data types, environment variables, statements, and keywords.

`Constant width bold`
    Shows commands or other text that should be typed literally by the user.

`Constant width italic`
    Shows text that should be replaced with user-supplied values or by values determined by context.

 This icon signifies a tip, suggestion, or general note.

 This icon indicates a warning or caution.

Using Code Examples

Supplemental material (code examples, exercises, etc.) is available for download at *http://www.family-science.net/electro_arduino.htm*.

This book is here to help you get your job done. In general, if example code is offered with this book, you may use it in your programs and documentation. You do not need to contact us for permission unless you're reproducing a significant portion of the code. For example, writing a program that uses several chunks of code from this book does not require permission. Selling or distributing a CD-ROM of examples from O'Reilly books does require permission. Answering a question by citing this book and quoting example code does not require permission. Incorporating a significant amount of example code from this book into your product's documentation does require permission.

We appreciate, but do not require, attribution. An attribution usually includes the title, author, publisher, and ISBN. For example: "*Make: Basic Arduino Projects* by Don Wilcher (Maker Media). Copyright 2014 Don Wilcher, 978-1-449-36066-5."

If you feel your use of code examples falls outside fair use or the permission given here, feel free to contact us at *bookpermissions@makermedia.com*.

Safari® Books Online

 Safari Books Online is an on-demand digital library that delivers expert content in both book and video form from the world's leading authors in technology and business.

With a subscription, you can read any page and watch any video from our library online. Read books on your cell phone and mobile devices. Access new titles before they are available for print, get exclusive access to manuscripts in development, and post feedback for the authors. Copy and paste code samples, organize your favorites, download chapters, bookmark key sections, create notes, print out pages, and benefit from tons of other time-saving features.

Maker Media has uploaded this book to the Safari Books Online service. To have full digital access to this book and others on similar topics from MAKE and other publishers, sign up for free at http://my.safaribooksonline.com (*http://my.safaribookson line.com/?portal=oreilly*).

How to Contact Us

Please address comments and questions concerning this book to the publisher:

MAKE
1005 Gravenstein Highway North
Sebastopol, CA 95472
800-998-9938 (in the United States or Canada)
707-829-0515 (international or local)
707-829-0104 (fax)

MAKE unites, inspires, informs, and entertains a growing community of resourceful people who undertake amazing projects in their backyards, basements, and garages. MAKE celebrates your right to tweak, hack, and bend any technology to your will. The MAKE audience continues to be a growing culture and community that believes in bettering ourselves, our environment, our educational system—our entire world. This is much more than an audience, it's a worldwide movement that Make is leading—we call it the Maker Movement.

For more information about MAKE, visit us online:

MAKE magazine: *http://makezine.com/magazine/*
Maker Faire: *http://makerfaire.com*
Makezine.com: *http://makezine.com*
Maker Shed: *http://makershed.com/*

We have a web page for this book, where we list errata, examples, and any additional information. You can access this page at *http://oreil.ly/basic-arduino*.

To comment or ask technical questions about this book, send email to *bookques tions@oreilly.com*.

Acknowledgments

I would like to thank Brian Jepson (Senior Editor) for believing in the book concept and allowing me to explore the Ultimate Microcontroller Pack in creative ways. Also, I would like to thank Patrick Di Justo (Editor) for pulling out the really cool projects from the original book proposal and coaching me to present them in fun and entertaining ways for young readers.

My final acknowledgment goes to my wife, Mattalene, who patiently worked with me on editing this book, keeping me on task with the writing/project builds, and reviewing the email revision messages from my editors. To my children, Tiana, D'Vonn, and D'Mar, thanks for being great kids while I worked on the book during family time.

The Trick Switch

Resistor-Capacitor Timing Basics

In electronics, sometimes we want to keep a device on for a certain amount of time even when an electrical switch is turned off. Ordinary pushbuttons used to turn electronic devices on and off can easily be operated by a timed delay switch. How awesome would it be to create such a device to delay turning off a simple LED? Such a gadget could be used to trick your friends, family, or even the local Makerspace when they see the LED staying on after the pushbutton has been released. With a few electronic components from the Ultimate Microcontroller Pack, you can make an LED (light-emitting diode) stay on for a few extra seconds when a pushbutton switch is turned off. Figure 1-1 shows an assembled Trick Switch. The electronic components required to build the Trick Switch are shown in the Parts List.

Parts List

- Arduino microcontroller
- SW1: mini pushbutton
- LED1: red LED
- C1: 100 uF electrolytic capacitor
- R1: 10K ohm resistor (brown, black, orange stripes)
- R2: 330 ohm resistor (orange, orange, brown stripes)
- Full-size clear breadboard

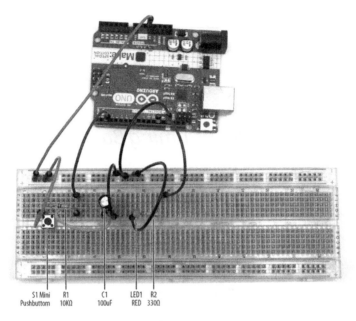

| S1 Mini | R1 | C1 | LED1 | R2 |
| Pushbuttom | 10KΩ | 100uF | RED | 330Ω |

Figure 1-1. *Trick Switch circuit built on a full-size clear breadboard (both the 100 uF electrolytic capacitor and red LED negative pins are wired to ground)*

Tech Note

You can create your own electrical circuits and test them using diagrams with an online simulator called Circuit Lab (*http://www.circuitlab.com*).

Let's Build a Trick Switch

When you press the pushbutton switch on this device, the LED turns on. The capacitor will begin storing electrical energy from the +5VDC power supply circuit of the Arduino. Releasing the pushbutton switch cuts off the flow of electricity from the source, but the energy stored in the capacitor keeps the Arduino running for a few extra seconds. The Arduino keeps the LED lit until the capacitor's stored energy is empty. You can build the Trick Switch using the electronic components from the Parts List and the Fritzing wiring diagram shown in Figure 1-2. Here are the steps required to build the electronic device:

1. From the Ultimate Microcontroller Pack, place the required parts on your workbench or lab tabletop.

2. Wire the electronic parts using the Fritzing wiring diagram of Figure 1-2 or the actual Trick Switch device shown in Figure 1-1.

3. Type the Pushbutton sketch shown in Example 1-1 into the Arduino text editor.

4. Upload the Pushbutton sketch to the Arduino.

5. Press the mini pushbutton for a moment. The red LED turns on. After one to two minutes, the red LED will turn off.

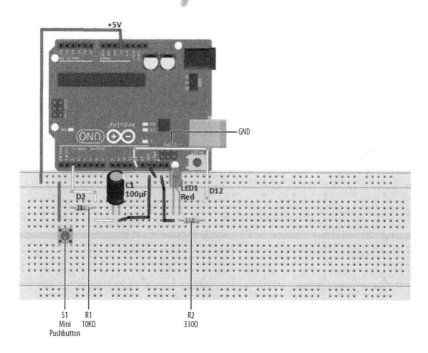

Figure 1-2. *Trick Switch Fritzing diagram*

Troubleshooting Tip

If the Trick Switch device doesn't work, check for incorrect resistor values, incorrect wiring, sketch typos, and improper orientation of polarized electronic components (the LED and capacitor).

Example 1-1. *Pushbutton sketch*

```
/*
  Pushbutton Sketch

Reads the capacitor voltage at digital pin 2 and turns on and off a light-
emitting diode (LED) connected to digital pin 12.


17 Nov 2012
```

by Don Wilcher

```
*/
// constants won't change; they're used here to
// set pin numbers:
const int buttonPin = 2;     // the number of the pushbutton pin
const int ledPin =  12;      // the number of the LED pin

// variables will change:
int buttonStatus = 0;        // variable for reading the pushbutton status

void setup() {
  // initialize the LED pin as an output:
  pinMode(ledPin, OUTPUT);
  // initialize the pushbutton pin as an input:
  pinMode(buttonPin, INPUT);
}

void loop(){
  // read the status of the pushbutton value:
  buttonStatus = digitalRead(buttonPin);

  // check if the pushbutton is pressed
  // if it is, the buttonEvent is HIGH:
  if (buttonStatus == HIGH) {
    // turn LED on:
    digitalWrite(ledPin, HIGH);
  }
  else {
    // turn LED off:
    digitalWrite(ledPin, LOW);
  }
}
```

 Tech Note
The ledPin value can be changed to 13 to operate the onboard LED.

Trick Switch with On/Off Indicators

In developing new products, electronics designers are always improving designs by
adding features and functions that excite the customer. The Trick Switch device you
built can be improved by adding an LED indicator. This LED indicates when the Trick
Switch timing cycle is done. Figure 1-3 shows you where to add a green LED to the
Trick Switch on the full-size clear breadboard.

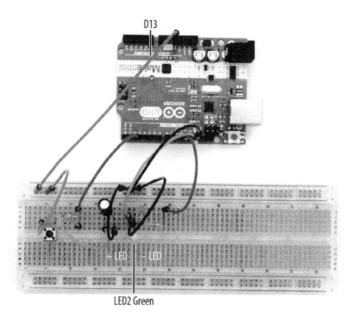

D13

LED2 Green

Figure 1-3. *Adding a green LED indicator to the Trick Switch circuit built on a full-size clear breadboard*

To complete the new product design, you need to make a few changes to the Push-button sketch. Modify the sketch using the code changes shown in Example 1-2.

Example 1-2. *Pushbutton sketch modified to include LED indicators*

```
// constants won't change; they're used here to
// set pin numbers:
const int buttonPin = 2;      // the number of the pushbutton pin
const int ledPin = 12;        // the number of the LED pin
const int ledPin13 = 13;      // onboard LED

void setup() {
  // initialize the LED pins as outputs:
  pinMode(ledPin, OUTPUT);
  pinMode(ledPin13, OUTPUT);
  // initialize the pushbutton pin as an input:
  pinMode(buttonPin, INPUT);
}

void loop(){
  // read the state of the pushbutton value:
  int buttonStatus;
  buttonStatus = digitalRead(buttonPin);

  // check if the pushbutton is pressed
  // if it is, the buttonStatus is HIGH:
  if (buttonStatus == HIGH) {
```

```
    // turn LED on:
    digitalWrite(ledPin, HIGH);
    // turn off onboard LED:
    digitalWrite(ledPin13,LOW);
  }
  else {
    // turn LED off:
    digitalWrite(ledPin, LOW);
    // turn on onboard LED:
    digitalWrite(ledPin13, HIGH);
  }
}
```

After you've saved the sketch changes and uploaded them to the Arduino, the green LED will turn on. When you press the mini pushbutton, the green LED will turn off, and the red LED will turn on. Pretty awesome stuff. Enjoy!

The block diagram in Figure 1-4 shows the electronic component blocks and the electrical signal flow for the Trick Switch. A Fritzing electronic circuit schematic diagram of the switch is shown in Figure 1-5. Electronic circuit schematic diagrams are used by electrical/electronic engineers to design and build cool electronic products for society.

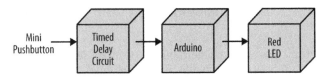

Figure 1-4. *Trick Switch block diagram*

Something to Think About

Try different resistor and capacitor values and see what happens. Can you detect any patterns? How can a small piezo buzzer be used with the Trick Switch?

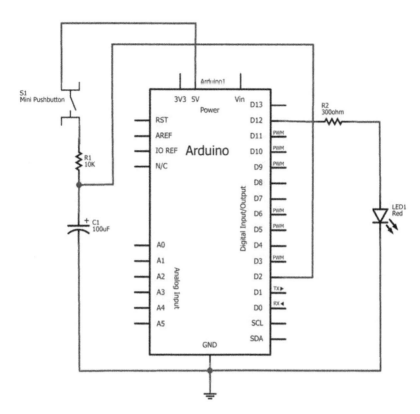

Figure 1-5. *Trick Switch circuit schematic diagram*

Sunrise-Sunset Light Switch | 2

Resistor-Capacitor Timing Basics

Designing and building new electronic devices is quite easy when you know the secret ingredient to rapid design. The technique is to take an existing electronic device and make a small change to it. For example, the Trick Switch project can easily be changed to a noncontact device by adding a sensor. With the right sensor, a hand wave can turn the LED on. The RC timing circuit wired to the sensor signals the Arduino to turn it off. In this chapter, you will build a light sensor circuit to turn on an LED with the wave of your hand and automatically turn it off. The Ultimate Microcontroller Pack has all the electronic parts you need to the build the project. Figure 2-1 shows the Sunrise-Sunset Light Switch device.

Parts List

- Arduino microcontroller
- SW1: mini pushbutton
- LED1: red LED
- LED2: green LED
- C1: 100 uF electrolytic capacitor
- R1: 10K ohm resistor (brown, black, orange stripes)
- R2: 330 ohm resistor (orange, orange, brown stripes)
- R3: photocell
- Full-size clear breadboard

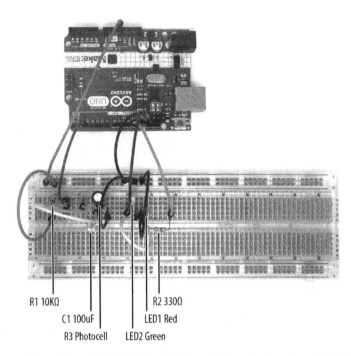

R1 10KΩ   R2 330Ω

C1 100uF   LED1 Red

R3 Photocell   LED2 Green

Figure 2-1. *Sunrise-Sunset Light Switch circuit built on a full-size clear breadboard (the 100 uF electrolytic capacitor and the red and green LED negative pins are wired to ground)*

Let's Build a Sunrise-Sunset Light Switch

You can build a Sunrise-Sunset Light Switch by modifying the Trick Switch device from Chapter 1. The main change you will make is to remove the mini pushbutton and replace it with a photocell. You will also add a green LED to pin D13 of the Arduino. Refer to the Parts List for all the electronic parts required for this project. Here are the steps required to build the electronic device:

1. From the Ultimate Microcontroller Pack, place the required parts on your work-bench or lab tabletop.

2. Wire the electronic parts using the Fritzing diagram of Figure 2-2 or the actual Sunrise-Sunset Light Switch device shown in Figure 2-1.

3. Type Example 2-1 into the Arduino IDE.

4. Upload the Sunrise-Sunset sketch to the Arduino. The green LED will be on.

5. Wave your hand over the photocell for a moment. The red LED turns on. After a few seconds, the red LED will turn off, and the green LED will turn on.

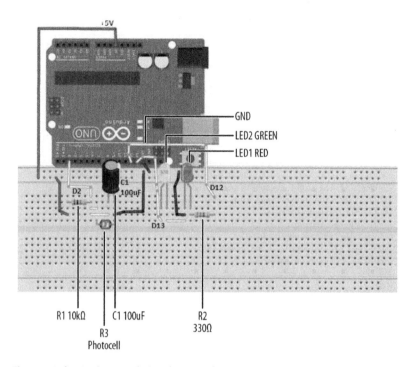

Figure 2-2. *Sunrise-Sunset Light Switch Fritzing diagram*

Example 2-1. *Sunrise-Sunset Light Switch sketch*

```
/*
  Sunrise-Sunset Light Switch

  Turns on and off a light-emitting diode (LED) connected to digital
  pins 12 and 13 after 10 to 20 seconds, by waving a hand over a photocell
  attached to pin 2.


  23 Nov 2012
  by Don Wilcher

*/


// constants won't change; they're used here to
// set pin numbers:
const int lightsensorPin = 2;  // the number of the light sensor pin
```

```
const int redledPin = 12;      // the number of the red LED pin
const int greenledPin13 = 13;  // onboard LED and green LED pin

// variables will change:
int sensorState = 0;           // variable for reading light sensor status

void setup() {
  // initialize the LED pins as outputs:
  pinMode(redledPin, OUTPUT);
  pinMode(greenledPin13, OUTPUT);
  // initialize the light sensor pin as an input:
  pinMode(lightsensorPin, INPUT);
}

void loop(){
  // read the state of the pushbutton value:
  sensorState = digitalRead(lightsensorPin);

  // check if the light sensor is activated
  // if it is, the sensorState is HIGH:
  if (sensorState == HIGH) {
    // turn red LED on:
    digitalWrite(redledPin, HIGH);
    // turn off onboard LED and green LED:
    digitalWrite(greenledPin13, LOW);
  }
  else {
    // turn red LED off:
    digitalWrite(redledPin, LOW);
    // turn on onboard LED and green LED;
    digitalWrite(greenledPin13, HIGH);
  }
}
```

Circuit Theory

The Sunrise-Sunset Light circuit operates like the Smart Switch, except you don't have to use a mini pushbutton to start the timing function. The mini pushbutton has instead been replaced with a light sensor called a photocell. A photocell is a variable resistor that changes its resistance based on the amount of light touching its surface. Light falling on a photocell will decrease its resistance value. No light will increase its resistance value. Figure 2-3 shows the resistor-capacitor (RC) timing circuit with a photocell variable resistor symbol.

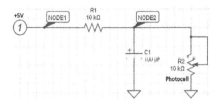

Figure 2-3. *RC timing circuit with a photocell (variable resistor)*

 Tech Note

Another type of variable resistor is a 3-pin electronic part known as a potentiometer. By rotating its shaft, the internal resistance value changes. Potentiometers are used in electronic products like radios and TVs to control the volume or sound level.

A photocell is a small electronic component with two leads protruding from a light-sensitive pellet. On top of the pellet is an etched series of squiggly lines. The lines etched on its surface are the resistance portion of the photocell. An actual photocell part can be seen in Figure 2-4.

Figure 2-4. *Photocell (a light-dependent resistor)*

If you know someone with a DMM (digital multimeter), have him attach your photocell to it. By waving your hand over the photocell, you will see the light-sensitive part change its resistance based on the amount of light touching it. This variable resistance feature will be used to turn on an LED. In the Sunrise-Sunset project (a light-activated switch), the green LED will be on first. Placing your hand over the photocell briefly will turn on the red LED. After the RC timing circuit has completed its charging-discharging cycle, the red LED turns off followed by the green LED turning on.

Sunrise-Sunset Detector with Serial Monitor

This Ultimate Microcontroller project demonstrates the power of electronic sensors to detect physical stimuli such as light, sound, and pressure. With a slight modification to the sketch, messages can scroll across a Serial Monitor. The Arduino IDE has a Serial Monitor for displaying the messages produced by the Arduino. You can access the Serial Monitor by following these two steps:

1. Move your mouse to the main toolbar of the Arduino IDE and click Tools.
2. Move the cursor, highlight "Serial Monitor," and click it.

The Serial Monitor will be displayed on your computer's screen. It's just that easy! The modifications to your sketch to display the messages "Sunrise" and "Sunset" on the Serial Monitor are shown in Example 2-2.

Example 2-2. *Sunrise Sunset Detector with Serial Monitor sketch*

```
const int lightsensorPin = 2;   // the number of the light sensor pin
const int redledPin = 12;       // the number of the red LED pin
const int greenledPin13 = 13;   // onboard LED and green LED pin

// variables will change:
int sensorState = 0;            // variable for reading light sensor status


void setup() {
  // initialize the LED pins as outputs:
  pinMode(redledPin, OUTPUT);
```

```
  pinMode(greenledPin13, OUTPUT);
  // initialize the light sensor pin as an input:
  pinMode(lightsensorPin, INPUT);
 // initialize serial communications at 9600 bps:
 Serial.begin(9600); // Add code instruction here!
}

void loop(){
  // read the state of the light sensor value:
  sensorState = digitalRead(lightsensorPin);

  // check if the light sensor is activated
  // if it is, the sensorState is HIGH:
  if (sensorState == HIGH) {
    // turn red LED on:
    digitalWrite(redledPin, HIGH);
    // turn off onboard LED and green LED:
    digitalWrite(greenledPin13, LOW);
    // display message
    Serial.println("Sunset\n"); // Add code instruction here!

  }
  else {
    // turn red LED off:
    digitalWrite(redledPin, LOW);
    // turn on onboard LED and green LED;
    digitalWrite(greenledPin13,HIGH);
    // display message
    Serial.println("Sunrise\n"); // Add code instruction here!
  }
}
```

With the modifications made to the original sketch, upload it to the Arduino and open the Serial Monitor. As you wave your hand over the photocell, you see the messages "Sunrise" (no hand over the sensor) and "Sunset" (hand over the sensor) displayed on the Serial Monitor. Figure 2-5 shows the two messages displayed on the Serial Monitor.

Experiment with the location of the Sunrise-Sunset detector to obtain the best circuit response. Enjoy!

The block diagram in Figure 2-6 shows the electronic component blocks and the electrical signal flow for the Sunrise-Sunset Light Switch. A Fritzing electronic circuit schematic diagram of the switch is shown in Figure 2-7. Electronic circuit schematic diagrams are used by electrical/electronic engineers to design and build cool electronic products for society.

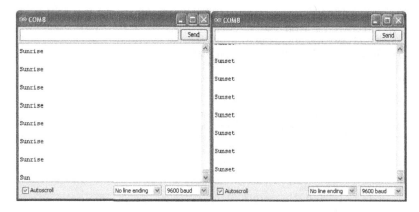

Figure 2-5. *Serial Monitor displaying "Sunset" and "Sunrise" messages*

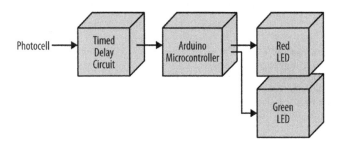

Figure 2-6. *Sunrise-Sunset Light Switch block diagram*

Tech Note
Always document your experiments and design changes in a lab notebook in case you develop that million dollar idea!

Something to Think About

How can the Serial Monitor display actual light sensor data?

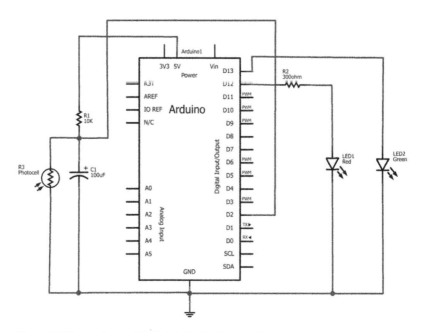

Figure 2-7. *Sunrise-Sunset Light Switch circuit schematic diagram*

Tilt Sensing Servo Motor Controller 3

Sensors allow people to operate consumer and industrial products using physical stimuli such as touch, sound, and motion. In Chapter 2, you controlled two LEDs with the wave of your hand; the light-activated switch used a photocell to detect the presence of your hand over the sensor.

In this chapter, we'll build a gadget to easily detect object orientation using a tilt control switch to control a servo motor. This is an awesome device to build and show your Maker smarts to family and friends. By rotating the tilt control switch in an upright position, you'll be able to operate a servo motor. Figure 3-1 shows the Tilt Sensing Servo Motor Controller.

Parts List

- Arduino microcontroller
- SW1: tilt control switch
- Jl: servo motor
- R1: 1K ohm resistor (brown, black, red stripes)
- Pair of alligator test leads or equivalent
- Full-size clear breadboard

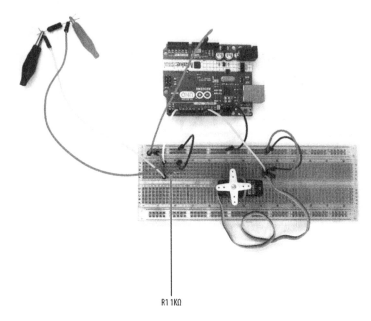

R1 1KΩ

Figure 3-1. *Tilt Sensing Servo Motor Controller built on a full-size clear breadboard*

Let's Build a Tilt Sensing Servo Motor Controller

You can control a servo motor's rotation direction through orientation detection using a tilt control switch. In this project, you will build a Tilt Sensing Servo Motor Controller. Refer to the Parts List for all the electronic components required for this project. Here are the steps used to build the electronic device:

1. From the Ultimate Microcontroller Pack, place the required parts on your workbench or lab tabletop.

2. Assemble the servo motor with the appropriate mechanical assembly attachment, as shown in Figure 3-2 (left).

3. Strip insulation from three ¼-inch solid wires and insert them into the servo motor's mini connector, as shown in Figure 3-2 (right).

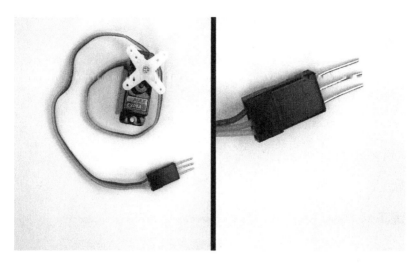

Figure 3-2. *Servo motor with mechanical assembly attachment and modified servo motor wire connector (left); close-up of modified servo motor wire connector (right)*

4. Place and secure the servo motor on the full-size clear breadboard with hookup wire, as shown in Figure 3-3.

5. Insert the modified servo motor wire connector into the full-size clear breadboard, as shown in Figure 3-4.

6. Wire the electronic parts using the Fritzing diagram of Figure 3-5, or the actual project shown in Figure 3-1.

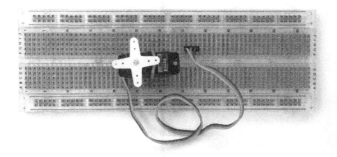

Figure 3-3. *Placing and securing the servo motor on the full-size clear breadboard*

Figure 3-4. *Modified servo motor wire connector inserted into the full-size clear breadboard*

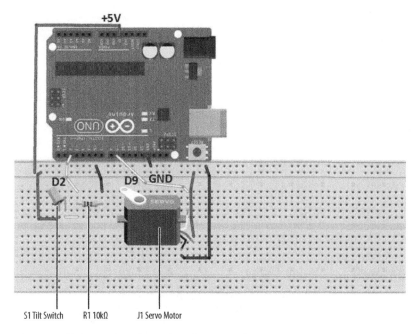

+5V

D2   D9   GND

S1 Tilt Switch   R1 10kΩ   J1 Servo Motor

Figure 3-5. *Tilt Sensing Servo Motor Controller Fritzing diagram*

Upload the Tilt Sensor Sketch

With the Tilt Sensing Servo Motor Controller circuit built on the full-size clear bread-board, it's time to upload the sketch:

1. Attach the Arduino to your computer using a USB cable.

2. Open the Arduino software and type Example 3-1 into the software's text editor.

3. Upload the sketch to the Arduino.

4. Rotate the tilt control switch back and forth. The servo motor will spin in the same direction as the tilt control switch orientation.

Troubleshooting Tip

Rotate the tilt control switch slowly and smoothly to get the best response from the servo motor.

Example 3-1. *Tilt Control Switch sketch*

```
/* This sketch controls a servo motor using a tilt control switch!
 *
 * 12 December 2012
 * by Don Wilcher
 *
 */

#include<Servo.h> // include Servo library
int inPin = 2;   // the tilt control switch is wired to Arduino D2 pin
int reading;     // the current reading from the input pin
Servo myservo; // create servo motor object

void setup()
{
    myservo.attach(9);       // attach servo motor to pin 9 of Arduino
    pinMode(inPin, INPUT); // make pin 2 an input

}

void loop()
{
    reading = digitalRead(inPin); // store digital data in variable
    if(reading == HIGH) {       // check digital data with target value
        myservo.write(180);     //  if digital data equals target value,
                                //  servo motor rotates 180 degrees
        delay(15);              //  wait 15ms for rotation
    }
    else {                      // if reading is not equal to target value,
        myservo.write(0);       // rotate servo motor to 0 degrees
        delay(15);              // wait 15ms for rotation
    }
}
```

A Simple Animatronic Controller Using the Serial Monitor

Tilt control switching can be used in a variety of human machine control and physical computing applications. In the theater, electromechanical puppets are often controlled using electronic sensors and servo motors. The tilt control switch project you built can be used as a simple animatronic controller. Example 3-2 shows an improved Arduino sketch that limits the servo motor rotation to 90° and adds Serial Monitor output for displaying information about the tilt control switch. Figure 3-6 shows the tilt control switch digital data.

Tech Note

Physical computing allows people to interact with objects using electronics, software, and sensors.

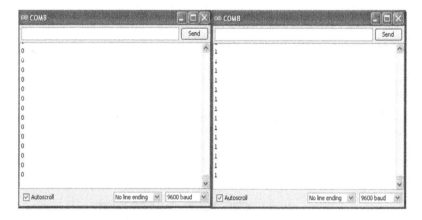

Figure 3-6. *Digital data from tilt control switch: open tilt control switch (left), closed tilt control switch (right)*

Example 3-2. *Tilt Control Switch with Serial Monitor*

```
/* This sketch controls a servo motor using a tilt control switch!
 * Serial Monitor displays digital data from Tilt Control Switch.
 *
 * 15 December 2012
 * by Don Wilcher
 *
 */

#include<Servo.h> // include Servo library
int inPin = 2;    //  the Arduino input pin tilt control switch is wired to D2
int reading;      //  the current reading from the input pin
Servo myservo; //   create servo motor object

void setup()
{
    myservo.attach(9);      // attach servo motor to pin 9 of Arduino
    pinMode(inPin, INPUT); //  make pin 2 an input
    Serial.begin(9600);    //  open communication port
}

void loop()
{
    reading = digitalRead(inPin);   // store digital data in variable
    if(reading == HIGH) {           // check it against target value (HIGH)

        myservo.write(90);          // if digital data equals target value,
                                    //  servo motor rotates 90 degrees
        Serial.println(reading);   //  print tilt control switch digital data
        delay(15);                 //   wait 15ms for rotation
    }
    else {                          // if it's not equal to target value...
```

```
        myservo.write(0);           //  rotate servo motor to 0 degrees
        Serial.println(reading); //  print tilt control switch digital data
        delay(15);                  //  wait 15ms for rotation
    }
}
```

Experiment with the different rotation angle values and observe the servo motor's behavior. Happy puppetry!

The block diagram in Figure 3-7 shows the electronic component blocks and the electrical signal flow for the Tilt Sensing Servo Motor Controller. A Fritzing electronic circuit schematic diagram of the controller is shown in Figure 3-8. Electronic circuit schematic diagrams are used by electrical/electronic engineers to design and build cool electronic products for society.

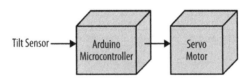

Figure 3-7. *Tilt Sensing Servo Motor Controller block diagram*

Tech Note
Always document your experiments and design changes in a lab notebook in case you develop that million dollar idea!

Observing the Switch's Behavior

If you have or know someone who has a DMM (digital multimeter), attach your tilt control switch to it. Set the DMM to read resistance. By changing the orientation of the tilt control switch, you will see the internal pins open and close based on the changing resistance value. A closed tilt control switch has a DMM resistance reading of zero ohms. An open tilt control switch has an infinite resistance value displayed on the DMM. The pins' switching conditions are used to rotate a servo motor. In the Tilt Sensing Servo Motor Controller project, directional movement will occur by orientation of the sensor circuit.

Note that an infinite resistance reading may be displayed as 0 or 1 on a DMM.

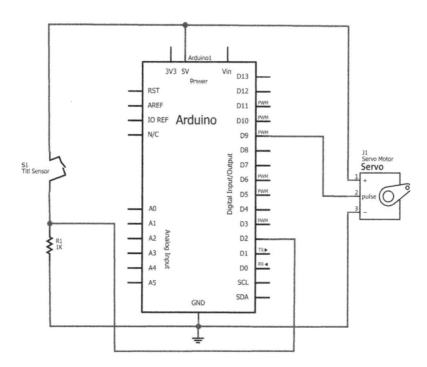

Figure 3-8. *Tilt Sensing Servo Motor Controller circuit schematic diagram: orange wire (D9), red wire (+5V), and brown wire (GND)*

Circuit Theory

A tilt control switch is an electrical device used to detect orientation. Like using a mini pushbutton and a light detector, a tilt control switch is another way to interact with and control the Arduino.

The tilt control switch is a pair of small metal balls that make contact with pins and close the circuit when the electrical device is held in an upright position. Figure 3-9 shows a typical tilt control switch. The tilt control switch can be wired to a resistor to make an orientation detection sensor circuit.

Figure 3-10 shows an orientation detection sensor circuit and its electrical operating conditions. The Arduino's D2 pin is wired to the 1KΩ resistor in order to receive either a zero or five volt control signal, based on the tilt control switch orientation. With the tilt control switch pins open, the voltage across the 1KΩ resistor is zero volts. When the switch pins are closed, the 1KΩ resistor has a five volt signal across it.

Figure 3-9. *Typical tilt control switch*

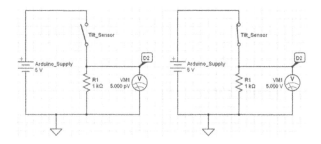

Figure 3-10. *Orientation detection sensor operation: open tilt control switch (left), closed tilt control switch (right)*

Something to Think About

How could the words "up" and "down" for the tilt sensor orientation be displayed on the Serial Monitor?

Twin LEDs

LEDs in Parallel

LEDs can be used to light objects or to alert the user about the operating conditions of a device. Light-emitting diodes are easy to use and they come in a variety of shapes, sizes, and colors, as shown in Figure 4-1. The Arduino has a dedicated tiny LED wired to pin D13. By uploading the Blink sketch to the Arduino, you can check its electrical operation. Unlike the ordinary light bulb, LEDs must be wired properly for them to work. In this chapter, you will learn how to wire two LEDs to the Arduino, as shown in the block diagram in Figure 4-2. The LED projects in this chapter will also use the Ultimate Microcontroller Pack's MakerShield. To add a little creativity to the LED projects, you will learn how to build an interactive toy using an ordinary piece of cardboard.

Parts List

- Arduino microcontroller
- MakerShield kit
- R1: 330 ohm resistor (orange, orange, brown stripes)
- R2: 330 ohm resistor (orange, orange, brown stripes)
- R3: 10K ohm potentiometer
- R4: photocell
- R5: 1K ohm resistor (brown, black, red stripes)

Figure 4-1. *Variety of LEDs*

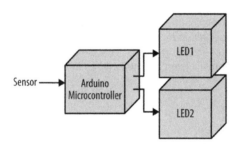

Figure 4-2. *Twin LEDs block diagram*

Circuit Theory

An LED is an electronic part that emits light when properly wired in an electric circuit. The LED has positive and negative leads protruding through a plastic body, as shown in Figure 4-1. You can use the Arduino in electronic projects to operate multiple LEDs. Figure 4-3 shows two LEDs wired to the Arduino D13 pin. The Arduino output pins are capable of providing 40 mA (milliamperes) of electrical current, sufficient to turn on two LED circuits wired in parallel.

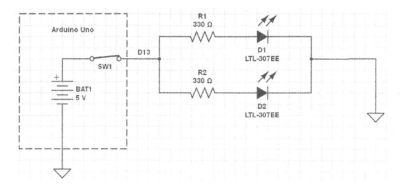

Figure 4-3. *Two LED circuits wired in parallel to the Arduino D13 pin; the arrows indicate the LEDs are on*

Twin LED Flasher

The circuit theory diagram shown in Figure 4-3 can easily be converted into a cool electronic gadget. You can build a Twin LED Flasher using an Arduino, two 330 ohm resistors, and LEDs, as shown in Figure 4-4. The Twin LED Flasher circuit schematic diagram is shown in Figure 4-5. To make the flasher device compact, you can build it on the MakerShield, as shown in Figure 4-6. Uploading the Blink sketch to the Arduino allows you to test the MakerShield and the Twin LED Flasher. The Blink sketch for the electronic flasher is shown in Example 4-1.

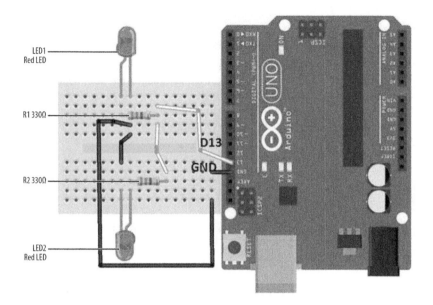

LED1
Red LED

R1 330Ω

D13

GND

R2 330Ω

LED2
Red LED

Figure 4-4. *Twin LED Flasher Fritzing diagram*

 Tech Note

The omega symbol (Ω) and the word ohm are used interchangeably.
For example, 10KΩ, 10K, and 10K ohm indicate the same value.

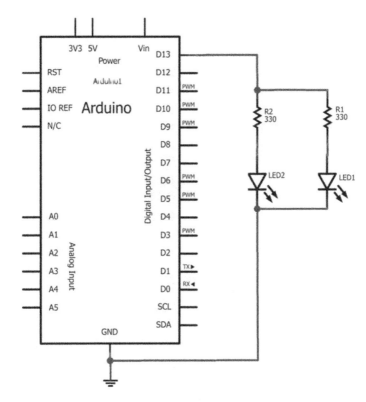

Figure 4-5. *Twin LED Flasher: LED1 and LED2 with 330 ohm resistors are wired in parallel to the Arduino D13 pin*

Tech Note

The Build a MakerShield guide (*http://bit.ly/lATbjJ*), part of Make: Projects, includes step-by-step directions for building your own prototyping shield.

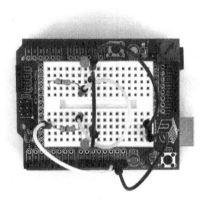

Figure 4-6. *MakerShield Twin LED Flasher*

Example 4-1. *Blink sketch*

```
/*
  Blink
  Turns on an LED on for one second, then off for one second, repeatedly.

  This example code is in the public domain.
 */

// Pin 13 has an LED connected on most Arduino boards.
// give it a name:
int led = 13;

// the setup routine runs once when you press reset:
void setup() {
  // initialize the digital pin as an output:
  pinMode(led, OUTPUT);
}

// the loop routine runs over and over again forever:
void loop() {
  digitalWrite(led, HIGH);   // turn the LED on (HIGH is the voltage level)
  delay(1000);               // wait for a second
  digitalWrite(led, LOW);    // turn the LED off by making the voltage LOW
  delay(1000);               // wait for a second
}
```

Build the Adjustable Twin LED Flasher

To make the Adjustable Twin LED Flasher, simply add a 10K ohm potentiometer to the device. The flash rate can be adjusted to make the on/off toggling slower or faster. The Fritzing diagram in Figure 4-7 along with the circuit schematic diagram shown in Figure 4-8 will allow you to build the Adjustable Twin LED Flasher. The MakerShield Adjustable Twin LED Flasher is shown in Figure 4-9 and the Adjustable Twin LED Flasher sketch is shown in Example 4-2.

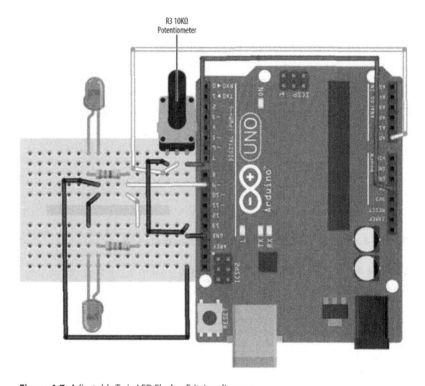

Figure 4-7. *Adjustable Twin LED Flasher Fritzing diagram*

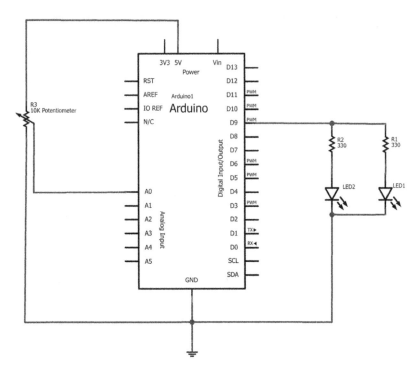

Figure 4-8. *Adjustable Twin LED Flasher circuit schematic diagram*

Example 4-2. *Adjustable Twin LED Flasher sketch*

```
/*
  Adjustable Twin LED Flasher
  Two LEDs will flash at a specified rate
  based on the 10K potentiometer setting.

  01 Jan 2013
  by Don Wilcher

 */

// Two LEDs with 330 ohm series resistors wired
// in parallel connected to pin 9.
int led = 9; // pin D9 assigned to led variable.

// A 10K potentiometer center pin wired to pin A0.
// One pin is wired to +5V with the other connected to GND.
int PotIn = A0; // pin A0 assigned to PotIn variable.

int Flash; // Flash variable to be used with "delay" instruction.

// the setup routine runs once when you press reset:
void setup() {
```

```
  // initialize the digital pin as an output:
  pinMode(led, OUTPUT);
  // initialize the analog pin as an input:
  pinMode(PotIn, INPUT);
}

// the loop routine runs over and over again forever:
void loop() {
  Flash =analogRead(PotIn);  // read 10K pot, store value in Flash variable
  digitalWrite(led, HIGH);  //  turn the LED on (HIGH voltage level = on)
  delay(Flash);             //    wait for a Flash time delay in seconds
  digitalWrite(led, LOW); //    turn the LED off by making the voltage LOW
  delay(Flash);             //    wait for a Flash time delay in seconds
}
```

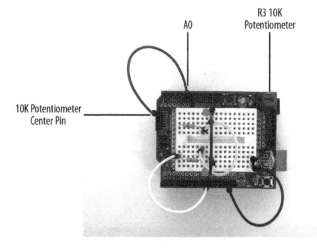

Figure 4-9. *MakerShield Adjustable Twin LED Flasher*

It's Alive! Build a FrankenBot Toy

You can build an interactive toy that responds to changing light levels by removing the 10KΩ potentiometer and adding a photocell wired to a 1KΩ resistor of the Adjustable Twin LED Flasher. Wiring a photocell to a 1KΩ resistor allows the Arduino to read light levels applied to pin A0. Figure 4-10 and Figure 4-11 show the Fritzing and circuit schematic diagrams for the Interactive Twin LED Flasher. The Maker-Shield Interactive Twin LED is shown in Figure 4-12.The photocell leads are bent down to allow FrankenBot's cardboard head to mount nicely on top of the Maker-Shield, as shown in Figure 4-13.

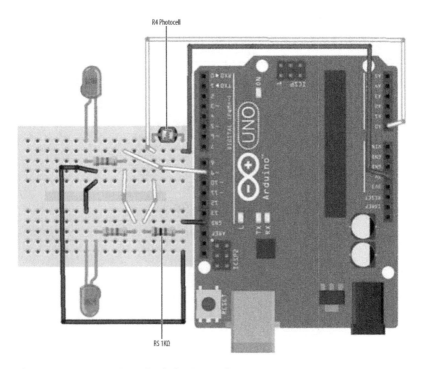

Figure 4-10. *Interactive Twin LED Flasher Fritzing diagram*

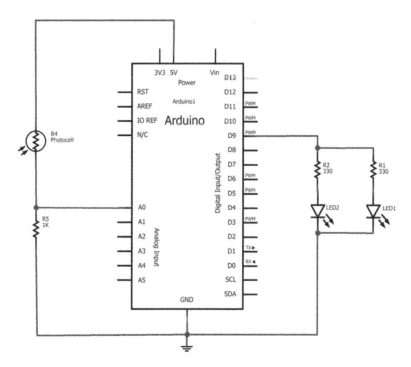

Figure 4-11. *Interactive Twin LED Flasher circuit schematic diagram*

The Adjustable Twin LED Flasher sketch shown in Example 4-2 is used to read the different light levels and change the Arduino's output flash rate.

Tech Note

Check out my YouTube clip (*http://youtu.be/ZcNf7hrvrNM*) to see the FrankenBot in action.

R5 1KΩ   R4 Photocell

Figure 4-12. *Makershield Interactive Twin LED Flasher*

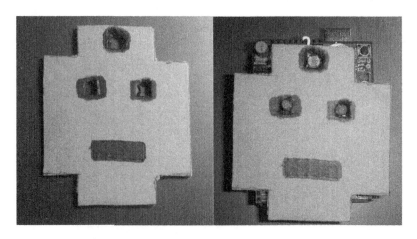

Figure 4-13. *FrankenBot: cut out opening for the photocell and LEDs to pass through cardboard FrankenBot head (left); mount cardboard Frankenbot head on top of MakerShield Interactive Twin LED Flasher (right)*

Tech Note

The FrankenBot head template can be found on my Arduino Downloads page (*http://www.family-science.net/electro_arduino.htm*).

With the Example 4-2 sketch uploaded to the Arduino, FrankenBot's LEDs will be flashing at a rate based on the surrounding lighting conditions. Place your hand over the photocell and watch the LEDs flash faster. Try different light sources and notice the effect on FrankenBot's LEDs. As always, record your tests and experiments in a lab notebook!

Troubleshooting Tip

Check your wiring, the placement of the electronic parts on the breadboard, and the photocell bent leads if the Twin LED Flasher is not working.

Something to Think About

Why are FrankenBot's LEDs affected by the changes in lighting conditions or sources?

The Opposite Switch 5

The Arduino NOT Logic Gate

Computers use electrical signals to make basic decisions. By hard-wiring electric circuits in specific ways, you can actually *see* simple logic decision operations at work. Ordinary electronic parts, like electrical switches, resistors, and LEDs can make AND, OR, and NOT logic gates when wired together properly.

The first computer logic decision circuit you will build is a NOT gate. You will also learn how the Opposite Switch works in electrical/electronic and digital computer circuits by building an Arduino NOT Logic Gate. Figure 5-1 shows the Arduino NOT Logic Gate (the Opposite Switch) device. The Ultimate Microcontroller Pack has all of the electronic parts to build this cool digital electronics device.

Parts List

- Arduino microcontroller
- MakerShield kit
- R1: 330 ohm resistor (orange, orange, brown stripes)
- R2: 330 ohm resistor (orange, orange, brown stripes)
- R3: 1KΩ resistor (brown, black, red stripes)
- SW1: pushbutton switch
- LED1: red LED
- LED2: green LED
- Battery1: 9VDC battery pack

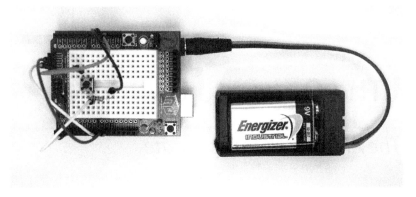

Figure 5-1. *The Arduino NOT Logic Gate*

Circuit Theory

A NOT Logic Gate turns a TRUE signal into a FALSE signal. Let's take the case of the ordinary household light switch: When you flip the light switch in your home UP, the light bulb turns on. Now, let's mount the house light switch upside down. When you send an UP signal to the switch, the light bulb will turn off. When you send a DOWN signal to the switch, the light bulb turns on. To illustrate this basic FALSE-TRUE operation, Figure 5-2 shows a simple NOT Logic Gate circuit you can build and experiment with, using a few electronic components from the Ultimate Microcontroller Pack. After wiring the NOT Logic Gate circuit on the breadboard, the red LED will be on. Pressing the pushbutton switch will turn the red LED off.

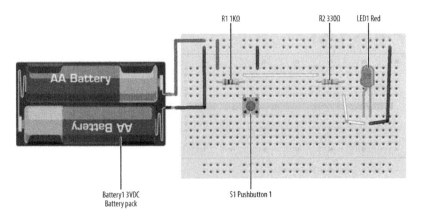

Figure 5-2. *A simple NOT Logic Gate Fritzing wiring diagram*

The Opposite Switch (aka the NOT Logic Gate)

In digital electronics, a special circuit symbol is used for the NOT Logic Gate consisting of a circle attached to the point of a triangle on its side. Figure 5-3 shows the digital electronics circuit symbol for the NOT Logic Gate.

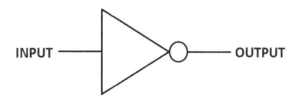

Figure 5-3. *The NOT Logic Gate circuit symbol*

In the Fritzing wiring diagram shown in Figure 5-2, the +3V battery provides the input binary data for the NOT Logic Gate and the electrical circuit output using an LED. With the switch open, the LED turns on. When the switch is closed, the LED turns off. Therefore, the purpose of the NOT Logic Gate circuit is to make a decision that is the opposite of the norm.

The NOT Logic Gate is traditionally used to invert a control signal used by smart machines like robots. To show the NOT Logic Gate's decision operation graphically, a truth table (TT) is used; this is shown in Figure 5-4. A truth table is used in digital electronics to show the operation of computer logic circuits in a simple data table. The input column is the digital data or information applied to the logic gate. The output column shows the logic gate's decision.

Tech Note
Another term for a NOT Logic Gate in digital electronics is "Inverter."

Input	Output
0	1
1	0

Figure 5-4. *The NOT Logic Gate truth table*

Build an Arduino NOT Logic Gate

We're going to add an Arduino microcontroller to Figure 5-2 to control two LEDs using a computer program or sketch. You will wire the pushbutton switch with a 1KΩ resistor to pin D2 and two LEDs (red and green). The LEDs will be attached to pins D8 and D9. Figure 5-5 shows the Fritzing wiring diagram for this project. The NOT Logic Gate can easily be built using the MakerShield prototyping board. This prototyping board makes the device portable so you can carry it in your shirt pocket or toolbox to demonstrate it to family, friends, and the local Makerspace. Refer to Figure 5-1 for the MakerShield NOT Logic Gate device.

Tech Note

The MakerShield prototyping board is an awesome tool to create cool electronic gadgets like the ones in this book or on the Makezine/Arduino projects website (*http://makezine.com/category/elec tronics/*).

Upload the Arduino NOT Logic Gate Sketch

With the Arduino NOT Logic Gate built on the MakerShield, it is time to upload the sketch. Example 5-1 operates the red and green LEDs using a pushbutton switch. Here are the steps you should follow:

1. Attach the Arduino to your computer using a USB cable.
2. Open the Arduino software and type Example 5-1 into the software's text editor.
3. Upload the sketch to the Arduino.
4. Press the mini pushbutton switch for a moment.

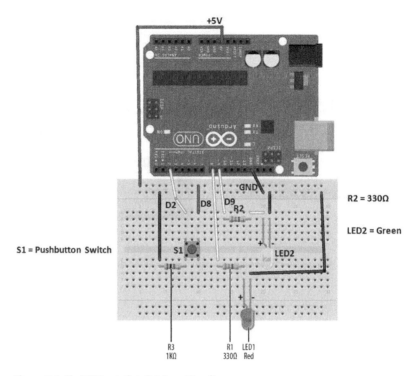

Figure 5-5. *The NOT Logic Gate Fritzing wiring diagram*

The Arduino NOT Logic Gate will turn the green LED on once the sketch has been uploaded to the microcontroller. Pressing the pushbutton switch will turn the green LED off and the red LED will be on. Figure 5-6 shows the Arduino NOT Logic Gate in operation. The green LED shows a TRUE output state when the pushbutton switch in not pressed. Pressing the pushbutton switch shows a FALSE output state by turning on the red LED. Also,the ! = in the Arduino sketch is the computer programming symbol for the logical NOT function.

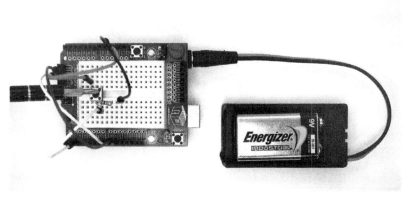

Figure 5-6. *The Arduino NOT Logic Gate: pressing the pushbutton switch turns on the red LED (FALSE output)*

Example 5-1. *The Arduino NOT Logic Gate sketch*

```
/*
  Arduino_NOT_Logic_Gate

  This sketch demonstrates the NOT(Inverter) Logic Gate operation.

  With the pushbutton switch not pressed (Logic LOW input), the green LED
  (Logic HIGH output indicator) is on and the red LED (Logic LOW output
  indicator) is off.
  Pressing the pushbutton turns the green LED off and the red LED on.

  11 September 2013
  by Don Wilcher

*/


// set pin numbers:
int buttonPin = 2;      // the number of the pushbutton pin
int LEDred =  8;        // pin number for the red LED
int LEDgreen = 9;       //   pin number for the green LED

// variables will change:
int buttonStatus = 0;           // variable for reading the pushbutton status

void setup() {
  // initialize the LED pins as outputs:
  pinMode(LEDred, OUTPUT);
  pinMode(LEDgreen, OUTPUT);
  // initialize the pushbutton pin as an input:
  pinMode(buttonPin, INPUT);
```

```
}

void loop(){
  // read the status of the pushbutton value:
  buttonStatus = digitalRead(buttonPin);

  // check if the pushbutton is not pressed
  //
  if (buttonStatus != HIGH) {
    // turn green LED on:
    digitalWrite(LEDgreen, HIGH);
    // turn red LED off:
    digitalWrite(LEDred, LOW);
  }
  else {
    // turn green LED off:
    digitalWrite(LEDgreen, LOW);
    // turn red LED ON:
  digitalWrite(LEDred, HIGH);
  }
}
```

After the sketch has been successfully uploaded to the Arduino microcontroller, the green LED will be on and the red LED off. Press the pushbutton switch to toggle the on/off states of the LEDs. To test the NOT Logic Gate, use the truth table shown in Figure 5-7.

Pushbutton Switch	LED1	LED2
Not Pressed	OFF	ON
Pressed	ON	OFF

Figure 5-7. *The Arduino NOT Logic Gate truth table*

Troubleshooting Tip

If the LEDs do not turn based on the truth table, check your electrical wiring and make sure the LEDs are properly oriented.

The block diagram in Figure 5-8 shows the building blocks and the electrical signal flow for the Arduino NOT Logic Gate. Circuit schematic diagrams are used by electrical engineers to quickly build cool electronic devices. The equivalent circuit schematic diagram for the Arduino AND Logic Gate is shown in Figure 5-9.

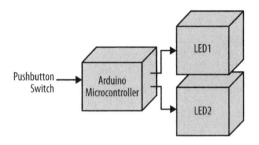

Figure 5-8. *The Arduino NOT Logic Gate block diagram*

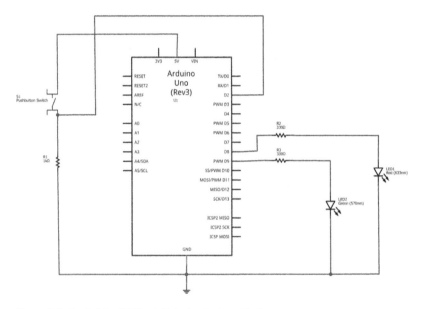

Figure 5-9. *The Arduino NOT Logic Gate circuit schematic diagram*

Something to Think About

How can a photocell be used to operate the Arduino NOT Logic Gate?

The AND Logic Gate

§

Two Pushbutton Switches in Series

The AND Logic Gate is another computer circuit used to make basic decisions with electrical input signals. Building the AND Logic Gate requires wiring two pushbutton switches to electrical contacts in a chain or series circuit. The AND Logic Gate's output decision is based on both parts of the input data (i.e., both pushbutton switches) being in the same state.

In this chapter, you will learn about the AND Logic Gate by building an Arduino-based AND Logic Gate using a photocell, a pushbutton switch, and an LED. Figure 6-1 shows the assembled Arduino AND Logic Gate. The Ultimate Microcontroller Pack has all of the electronic parts to build this interactive tutorial device.

Parts List

- Arduino microcontroller
- MakerShield kit
- S1: pushbutton switch
- S2: pushbutton switch
- R1: 1KΩ resistor (brown, black, red stripes)
- R2: 10KΩ potentiometer (for basic AND Logic Gate circuit)
- R2: photocell (for Arduino AND Logic Gate circuit)
- R3: 1KΩ resistor (brown, black, red stripes) for basic AND Logic Gate circuit
- R3: 10KΩ resistor (brown, black, orange stripes) for Arduino AND Logic Gate circuit
- R4: 330Ω resistor (orange, orange, brown stripes)

- LED1: green LED (for Arduino AND Logic Gate circuit)
- LED1: red LED (for basic AND Logic Gate circuit)
- Battery1: 3VDC battery pack

Figure 6-1. *The assembled Arduino AND Logic Gate*

Tech Note

In digital electronics, a TRUE state is when a data bit is set to 1, or an output pin is set to +5 volts, or a switch is closed. A FALSE state is when a data bit equals 0, or an output pin is set to 0 volts, or a switch is open.

Circuit Theory

The AND Logic Gate is a computer circuit that outputs TRUE if the two pieces of input data have the same state as each other. Digital electronics have two binary states: TRUE or FALSE. The electric circuits used in digital electronics to build computer logic gates will either be closed or open representing a TRUE or FALSE state.

The schematic diagram in Figure 6-2 shows a basic AND Logic Gate electric circuit in two different states. On the left, we see the circuit with both pushbutton switches open. In this state, the output for both pushbutton switches is FALSE, and the LED will be off. On the right, we see the circuit when both switches are closed. In this state, the output is TRUE, and the LED will be on.

The operation of the AND Logic Gate can easily be programmed into an Arduino microcontroller, as shown in Figure 6-1. Creating an Arduino AND Logic Gate requires a few basic electronic components found in the Ulitmate Microcontroller Pack. The logical AND operator is part of the Arduino sketch library. The Arduino AND Logic Gate will turn on an LED when two inputs are *both* TRUE and are logical HIGH. "Why Use an Arduino Microcontroller to Build an AND Logic Gate?" on page 54 explains the advantages of using a smart chip to create logical gate functions.

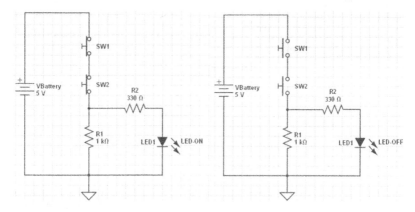

Figure 6-2. *Circuit schematic diagram showing an AND Logic Gate controlling an LED*

Tech Note

When electrical/electronic components are connected with the output of one component becoming the input of the next, we say that they have made a *series* circuit.

Figure 6-3 shows the Fritzing wiring diagram.

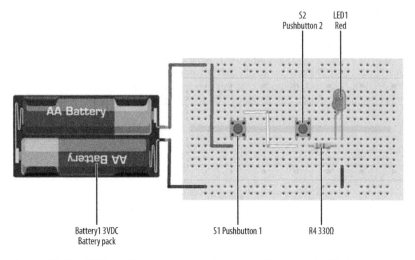

Battery1 3VDC
Battery pack

S2
Pushbutton 2

LED1
Red

AA Battery

AA Battery

S1 Pushbutton 1

R4 330Ω

Figure 6-3. *The AND Logic Gate Fritzing wiring diagram; the flat side of the LED is the negative pin*

Just like the NOT Logic Gate discussed in Chapter 5, the AND Logic Gate has a special circuit symbol, shown in Figure 6-4. The truth table (TT) shows the logic gate operation. Figure 6-5 is an AND Logic Gate TT.

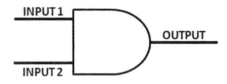

INPUT 1

OUTPUT

INPUT 2

Figure 6-4. *The AND Logic Gate circuit symbol*

INPUT1	INPUT2	INPUT3
OFF	OFF	OFF
OFF	ON	OFF
ON	OFF	OFF
ON	ON	ON

Figure 6-5. *The AND Logic Gate truth table*

Why Use an Arduino Microcontroller to Build an AND Logic Gate?

The AND Logic Gate is a simple electrical circuit where a string of pushbutton switches are wired in series. You can wire it up yourself right now, and get the same results without an Arduino. So the question "Why use an Arduino microcontroller to build an AND Logic Gate?" is very important.

The Arduino microcontroller, like all microcontrollers, has a bunch of programming instructions that are essentially built around logic operators (AND, exclusive OR [XOR], OR, and NOT [Complement(CPL)]). You can use these logic operators to make decisons based on the state of the pushbutton switches (open or closed) and using the appropriate logic operator. As you'll see, you'll actually be able to make the same circuit behave differently, simply by changing the Arduino code! Aren't computers wonderful?

The Arduino AND Logic Gate

We're going to replicate the AND gate using an Arduino. To make the circuit interesting, we're going to use a photocell to replace one of the pushbutton switches from Figure 6-1. In Figure 6-6, we see the Arduino AND Logic Gate, with the LED turned off. Placing a piece of tape over the photocell (to allow no light into it, thus simulating nighttime) will make input pin D4 TRUE for the Arduino microcontroller. Pressing the pushbutton switch allows input pin D3 to become TRUE. With both inputs TRUE, the green LED turns on, as shown in Figure 6-7.

Figure 6-8 shows the Fritzing wiring diagram to use for building the Arduino AND Logic Gate. As shown in Figures 6-1, 6-7, and 6-8, you can build the Arduino AND Logic Gate on a MakerShield. The MakerShield makes the project portable so that you can carry it to show family and friends the basic logic gate used in computers, cell phones, robotics, and other smart electronic devices.

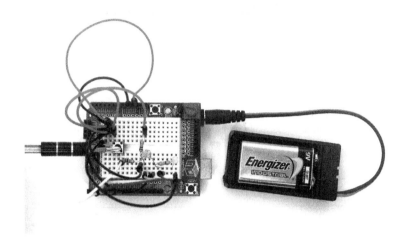

Figure 6-6. *The Arduino AND Logic Gate with LED turned off*

Figure 6-7. *The Arduino AND Logic Gate with LED turned on*

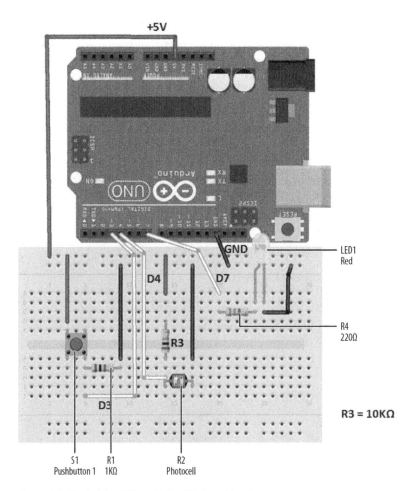

Figure 6-8. *The Arduino AND Logic Gate Fritzing wiring diagram*

 Tech Note
The flat side of an LED is the negative pin.

Upload the Arduino AND Logic Gate Sketch

With the Arduino AND Logic Gate built on the MakerShield, it is time to upload the sketch. Example 6-1 operates the green LED using a pushbutton switch and a photocell. Here are the steps you'll need to follow:

1. Attach the Arduino to your computer using a USB cable.

2. Open the Arduino software and type Example 6-1 into the software's text editor.

3. Upload the sketch to the Arduino.

4. Press the mini pushbutton switch for a moment.

The Arduino AND Logic Gate will turn on the LED when the photocell is covered *and* the pushbutton switch is pressed. Releasing the pushbutton switch, or placing a light on the photocell, turns the LED off, because the AND condition (in which both switches are closed) no longer exists.

The Arduino does this by using the && operator in the *if* statement. && is the computer programming symbol for the logical AND function.

Example 6-1. *The Arduino AND Logic Gate sketch*

```
/*
  The Arduino AND Logic Gate

  Turns on an LED connected to digital
  pin 7, when pressing a pushbutton switch and covering a photocell
  attached to pins 3 and 4.

27 Jan 2013
Revised 4 September 2013
by Don Wilcher

 */


// constants won't change; they're used here to
// set pin numbers:
int B = 3;      // the number of the B pushbutton pin
int A = 4;      //  the number of the A pushbutton pin

const int Cout =  7;      // the number of the LED pin

// variables will change:
int AStatus = 0;          // variable for reading the A pushbutton status
int BStatus = 0;
void setup() {
  // initialize the LED pin as an output:
  pinMode(Cout, OUTPUT);
  // initialize the pushbutton pins as inputs:
  pinMode(B, INPUT);
  pinMode(A, INPUT);
}

void loop(){
  // read the state of the pushbutton value:
  AStatus = digitalRead(A);
  BStatus = digitalRead(B);
```

```
// check if the pushbuttons are pressed
// if it is, the buttonStatus is HIGH:
if (AStatus == HIGH && BStatus ==HIGH) {
  // turn LED on:
  digitalWrite(Cout, HIGH);
  }
else {
  // turn LED off:
  digitalWrite(Cout, LOW);
  }
}
```

After uploading the Arduino AND Gate Logic sketch to the Arduino microcontroller, the green LED is turned off. As discussed earlier, pressing the pushbutton switch and covering the photocell will turn on the green LED. If either input device is FALSE (logic LOW), the LED turns off. To completely test the Arduino AND Logic Gate's operation, remember to use the TT shown in Figure 6-5.

The block diagram in Figure 6-9 shows the building blocks and the electrical signal flow for the Arduino AND Logic Gate. Circuit schematic diagrams are used by electrical engineers to quickly build cool electronic devices. The equivalent circuit schematic diagram for the Arduino AND Logic Gate is shown in Figure 6-10.

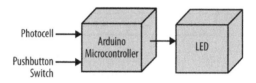

Figure 6-9. *The Arduino AND Logic Gate block diagram*

Troubleshooting Tip

If you don't see the green LED turn on with the proper input logic, recheck the logic gate's electrical wiring. Also, check to see if the green LED's negative pin is wired to GND on the breadboard.

Something to Think About

How can the photocell be wired to analog A0 and be used as a digital input pin?

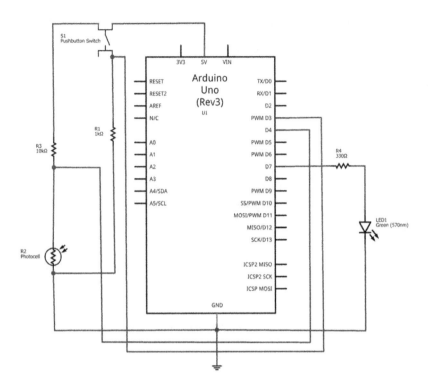

Figure 6-10. *The Arduino AND Logic Gate circuit schematic diagram*

The OR Logic Gate 7

Two Pushbutton Switches in Parallel

The OR Logic Gate is the final basic computer circuit used to make simple decisions with electrical input signals. The OR Logic Gate is different from the AND circuit because the two pushbutton switches are connected in *parallel* (instead of in *series* as with the AND circuit). The OR Logic Gate's output decision is based on either one *or* the other input being TRUE. In this chapter, you will learn about the OR Logic Gate by building an Arduino OR Logic Gate using one pushbutton switch and a photocell. Figure 7-1 shows the assembled Arduino OR Logic Gate. The Ultimate Microcontroller Pack has all of the electronic parts to build this cool digital computer circuit.

Parts List

- Arduino microcontroller
- MakerShield kit
- S1: pushbutton switch
- S2: pushbutton switch
- R1: 1KΩ resistor (brown, black, red stripes)
- R1: 330Ω resistor (orange, orange, brown stripes) for basic OR Logic Gate circuit
- R2: photocell
- R3: 1KΩ resistor (brown, black, red stripes)
- R4: 330Ω resistor (orange, orange, brown stripes) for Arduino OR Logic Gate
- LED1: red LED for basic OR Logic Gate Circuit (green LED for Arduino OR Logic Gate)

Figure 7-1. *The assembled Arduino OR Logic Gate*

Tech Note

When two or more electrical/electronic components are connected across one voltage source via separate paths, this is called a parallel circuit. In a parallel circuit, there are different ways that electricity can flow.

Circuit Theory

The OR Logic Gate is another computer circuit providing a TRUE output based on at least one input data value having a closed state. A basic OR Logic Gate electric circuit schematic diagram is shown in Figure 7-2. The circuit schematic diagram on the left shows one pushbutton switch closed and the other one open. The output for this pushbutton switch combination is TRUE, indicated by the LED being on. When both switches are open, as shown in the right circuit schematic diagram, the LED will be off. In summary, if at least one of the pushbutton switches is TRUE, the OR Logic Gate's output will be TRUE. Therefore, the OR Logic Gate provides a TRUE output when *either* input is TRUE. This is different from the AND gate, which requires *both* inputs to be TRUE. To experiment with a basic OR Logic Gate circuit, the Fritzing wiring diagram shown in Figure 7-3 can be built on a breadboard.

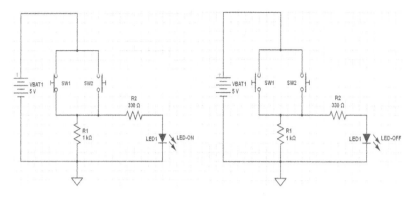

Figure 7-2. *Circuit schematic diagram for the OR Logic Gate controlling an LED*

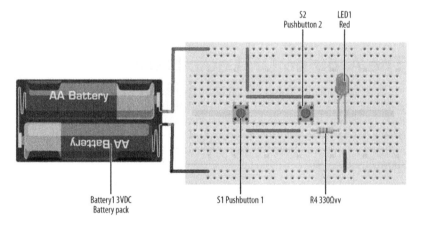

Figure 7-3. *A basic OR Logic Gate circuit Fritzing wiring diagram*

Tech Note

If you need an OR Logic Gate for an electronics project, simply wire two or more switches in parallel.

Just like the other logic gates discussed in Chapter 5 and Chapter 6, the OR Logic Gate has a special circuit symbol as well, shown in Figure 7-4. The truth table (TT) shows the logic gate operation. Figure 7-5 is an OR Logic Gate TT.

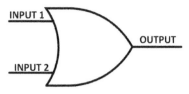

Figure 7-4. *The OR Logic Gate circuit symbol*

INPUT1	INPUT2	INPUT3
OFF	OFF	OFF
OFF	ON	ON
ON	OFF	ON
ON	ON	ON

Figure 7-5. *The OR Logic Gate Truth Table*

The Arduino OR Logic Gate

You can build a digital computer OR Logic Gate circuit using the Arduino micro-controller and a few electronic components from the Ultimate Microcontroller Pack. The green LED turns on when either the pushbutton switch OR the photocell is TRUE. You can easily build the logic circuit using the Fritzing wiring diagram shown in Figure 7-6. You can build this basic digital computer circuit on MakerShield, as shown in Figure 7-1.

Did you notice that the Fritzing wiring diagram looks like the AND Logic Gate circuit of Chapter 6? That's because it *is*. The cool thing about using an Arduino (or any other computer, really) is that often you can use the same physical circuit and make it do different things, simply by changing the computer code. In this case, either pressing the pushbutton switch OR placing your hand over the photocell will turn on the green LED.

This cool gadget can become an automatic LED night light. If your home loses power because of an electrical storm or the area substation is not operating, this device can function as an automatic light source. The photocell is electrically wired to detect darkness. When night falls (or when the power fails), the signal at pin D4 becomes TRUE, and the Arduino microcontroller turns on the green LED, as in Figure 7-7. Or, if you just want to turn the light on when it isn't dark out, you can just hit the push-button switch. This makes the signal at pin D3 TRUE, which again causes the Arduino microcontroller to turn on the green LED.

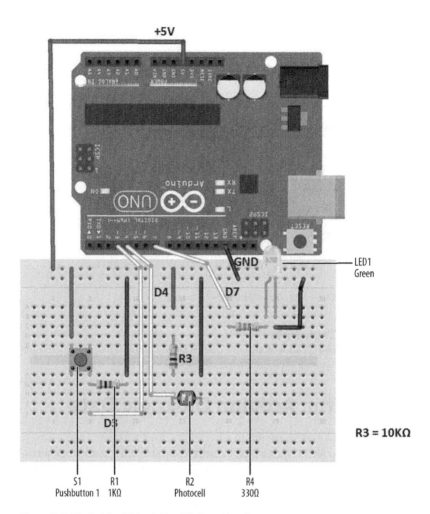

+5V

D4

GND

D7

LED1
Green

R3

D3

R3 = 10KΩ

S1
Pushbutton 1

R1
1KΩ

R2
Photocell

R4
330Ω

Figure 7-6. *The Arduino OR Logic Gate Fritzing wiring diagram*

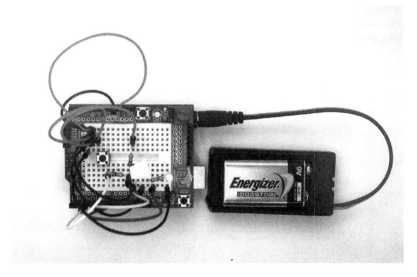

Figure 7-7. *The LED is on: the photocell is covered with tape*

Tech Note
The incandescent light bulb is slowly being replaced by LEDs because of their low power consumption and long life.

Upload the Arduino OR Logic Gate Sketch

With the Arduino AND Logic Gate built on the MakerShield, it is time to upload the sketch. Example 7-1 operates the green LED using a pushbutton switch and a photocell. Here are the steps you'll need to take:

1. Attach the Arduino to your computer using a USB cable.

2. Open the Arduino software and type Example 7-1 into the software's text editor.

3. Upload the sketch to the Arduino.

4. Press the mini pushbutton switch for a moment.

The Arduino OR Logic Gate will turn on the LED when the photocell is covered or the pushbutton switch is pressed. Releasing the pushbutton switch or placing a light on the photocell turns off the LED, because an OR condition (in which either switch is closed) no longer exists.

The Arduino does this by using the || operator in the *if* statement. || is the computer programming symbol for the logical OR function.

Example 7-1. *The Arduino OR Logic Gate sketch*

```
/*
  The Arduino OR Logic Gate

  Turns on an LED connected to digital
  pin 7, when pressing either a pushbutton switch or covering a photocell
  attached to pins 3 and 4.

  27 Jan 2013
  Revised 4 September 2013
  by Don Wilcher



 */

// constants won't change; they're used here to
// set pin numbers:
int B = 3;      // the number of the B pushbutton pin
int A = 4;      //  the number of the A pushbutton pin

const int Cout =  7;      // the number of the LED pin

// variables will change:
int AStatus = 0;          // variable for reading the A pushbutton status
int BStatus = 0;
void setup() {
  // initialize the LED pin as an output:
  pinMode(Cout, OUTPUT);
  // initialize the pushbutton pins as inputs:
  pinMode(B, INPUT);
  pinMode(A, INPUT);
}

void loop(){
  // read the state of the pushbutton value:
  AStatus = digitalRead(A);
  BStatus = digitalRead(B);

  // check if the pushbuttons are pressed
  // if it is, the buttonStatus is HIGH:
  if (AStatus == HIGH || BStatus ==HIGH) {
    // turn LED on:
    digitalWrite(Cout, HIGH);
   }
  else {
    // turn LED off:
    digitalWrite(Cout, LOW);
   }
}
```

After uploading the Arduino OR Logic Gate sketch to the Arduino microcontroller, the green LED is off. Pressing the pushbutton switch or placing your hand over the photocell will turn on the green LED. To completely test the Arduino OR Logic Gate's operation, remember to use the TT shown in Figure 7-5.

The block diagram in Figure 7-8 shows the building blocks and the electrical signal flow for the Arduino OR Logic Gate. Circuit schematic diagrams are used by electrical engineers to quickly build cool electronic devices. The equivalent circuit schematic diagram for the Arduino OR Logic Gate is shown in Figure 7-9.

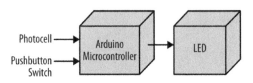

Figure 7-8. *The Arduino OR Logic Gate block diagram*

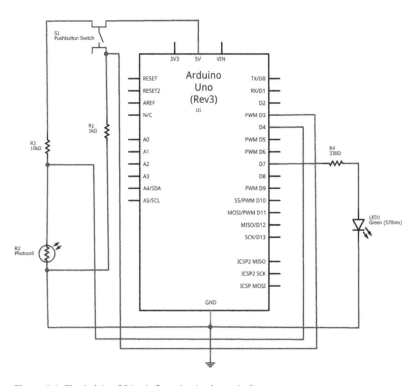

Figure 7-9. *The Arduino OR Logic Gate circuit schematic diagram*

Troubleshooting Tip

If the LED doesn't light up after uploading the sketch to the Arduino microcontroller, check to see if it is connected to the correct output pin. Also, check to see if the LED has been connected in the proper orientation, with the short LED wire connected to GND.

Something to Think About

How can the Arduino OR Logic Gate be used to flash the green LED when the push-button switch or the photocell is TRUE?

Tilt Flasher

8

Up-Down Sensor

In Chapter 4, the FrankenBot toy illustrated the method of flashing two LEDs wired in parallel. The Arduino microcontroller made it easy to change the flash rate of two LEDs using a 10KΩ potentiometer. How cool would it be to control an LED's flash rate by moving an electronic box in an up-down motion? This project is a small window into the inner workings of gesture controls used to operate video games, robots, and electronic toys using simple body motions. Now we're going to build and use orientation control, in the form of a tilt switch, to operate an Arduino flasher. The parts for the project consist of a tilt control switch, three resistors, and two LEDs. Figure 8-1 shows the Up-Down Sensor block diagram. You will find all of these electronic parts in the Ultimate Microcontroller Pack.

Parts List

- Arduino microcontroller
- MakerShield kit
- S1: tilt control switch
- R1: 1KΩ resistor (brown, black, red stripes)
- R2: 330Ω resistor (orange, orange, brown stripes)
- R3: 330Ω resistor (orange, orange, brown stripes)
- LED1: green LED
- LED2: red LED

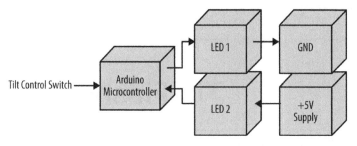

Figure 8-1. *The Up-Down Sensor block diagram*

 Tech Note

A tilt control switch is sometimes called a tilt sensor.

Circuit Theory

As shown in Figure 8-1, the LED wiring is quite different from previous projects, because an electrical ground and a +5V battery are used to individually operate them. The idea behind this wiring technique is to allow one LED to be on at all times. The circuit schematic diagram in Figure 8-2 on the left shows LED1 on while LED2 is off. The DPDT (double pole, double throw) switch upper contacts are open with the bottom contacts closed. The closed contacts allow current from the battery (+5V) to flow through the LED (LED1) turning it on. The open contacts turn off LED2 because no battery current is flowing through it. The Arduino microcontroller will use this DPDT switching method for the Up-Down Sensor project to operate two LED orientation indicators.

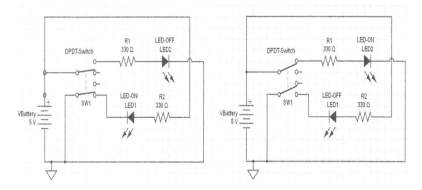

Figure 8-2. *Circuit schematic diagram for a DPDT switch toggling two LEDs*

Tech Note
Current is the movement or flow of electric charge.

The Up-Down Sensor

A simple Arduino microcontroller flasher can easily be turned into an Up-Down Sensor by adding a tilt control switch. As shown in Figure 8-3, the tilt control sensor is mounted on the MakerShield. When the tilt control switch is in the horizontal position, both the red and green LEDs will flash. Placing the MakerShield on its side will rotate the tilt control switch to vertical. The red LED will turn off and the green LED will be on but not flashing. Rotating the MakerShield back to the horizontal position causes the red and green LEDs to resume flashing. You can use either the Fritzing diagram shown in Figure 8-4 or the circuit schematic diagram in Figure 8-5 to build the Up-Down Sensor.

Tech Note
Placing another tilt control switch on the MakerShield in a vertical position can provide back and forth detection for robotics projects.

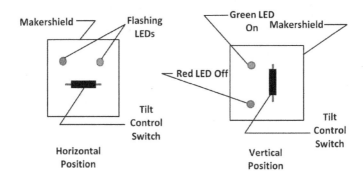

Figure 8-3. *The Up-Down Sensor concept diagram*

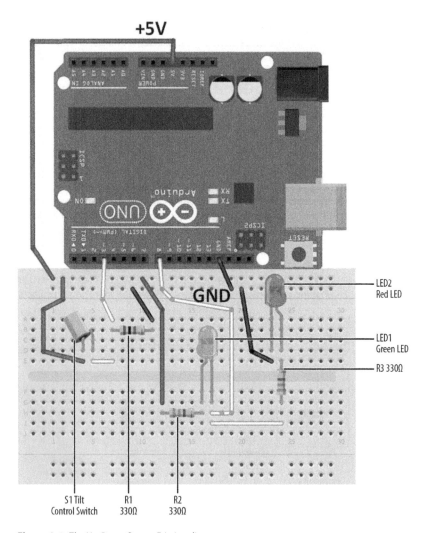

+5V

GND

LED2
Red LED

LED1
Green LED

R3 330Ω

S1 Tilt
Control Switch

R1
330Ω

R2
330Ω

Figure 8-4. *The Up-Down Sensor Fritzing diagram*

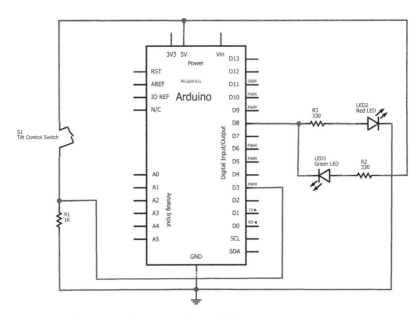

Figure 8-5. *The Up-Down Sensor circuit schematic diagram*

You can build the Up-Down Sensor on a MakerShield, as shown in Figure 8-6. The MakerShield allows you to carry it in a shirt pocket, computer bag, or purse for convenience. Example 8-1 can be uploaded to the Arduino after entering the code into the IDE's text editor screen.

Example 8-1. *Up-Down Sensor sketch*

```
/*
  Up-Down Sensor with Flashing LEDs

  Flashes green and red LEDs at pin 8 when the tilt control
  switch attached to pin 3 is tilted. The green LED wired to
  pin 8 turns turns solid when no tilt condition is detected.

  05 Feb 2013
  Don Wilcher

 */

// constants won't change; they're used here to
// set pin numbers:
const int tiltPin = 3;     // the number of the tilt control switch pin
const int ledPin = 8;      // the number of the LED pin

// variables will change:
int tiltState = 0;         // variable for tilt control switch status

void setup() {
```

```
  pinMode(ledPin, OUTPUT);
  // initialize the tilt control switch pin as an input:
  pinMode(tiltPin, INPUT);
}

void loop(){
  // read the state of the tilt control switch value:
  tiltState = digitalRead(tiltPin);

  // check if the tilt control switch contacts are closed
  // if they are, the tiltState is HIGH:
  if (tiltState == HIGH) {
    // turn Red LED on;
    digitalWrite(ledPin, HIGH);
    // wait 100ms:
    delay(100);
    // turn LED off:
    digitalWrite(ledPin,LOW);
    // wait 100ms:
    delay(100);
  }
  else {
    // turn LED off:
    digitalWrite(ledPin, LOW);
  }
}
```

After uploading the Up-Down Sensor sketch to the Arduino microcontroller, orient the MakerShield so that the tilt control switch is horizontal. The red and green LEDs should be flashing. Next, rotate the MakerShield onto its side. The red LED will be off and the green LED will be on. Experiment with different MakerShield orientations and notice the response of the LEDs. Remember to record your observations in your lab notebook!

Troubleshooting Tip

If the LEDs don't turn on after uploading the sketch to the Arduino microcontroller, check that the correct output pin is being used. Also, check to see that the LEDs are wired up properly, with the short wires connected to GND.

Something to Think About

How can the Up-Down Sensor be used to turn a small electric DC motor on and off?

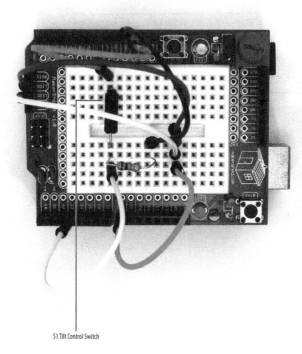

S1 Tilt Control Switch

Figure 8-6. *The Up-Down Sensor built on a MakerShield*

Multicolor RGB Flasher

9

Free Running Switcher

In past projects, one or two LEDs, usually red and/or green, were used as visual indicators, letting us know that the Arduino had completed a task or operation. But why limit ourselves to red and green? There is a type of LED that has three different colors all in the same package. An RGB LED has three light-emitting diodes inside of it: one red, one green, and one blue.

In this chapter, you will learn how to use a multicolor LED by building a simple RGB Flasher. The parts you will use for the RGB Flasher are one fixed resistor, a multicolor LED, and an Arduino microcontroller. The RGB Flasher will be built using the handy MakerShield. The Ultimate Microcontroller Pack has all of the project parts you need. Figure 9-1 shows the RGB Flasher block diagram.

Parts List

- Arduino microcontroller
- MakerShield kit
- R1: 330Ω resistor (orange, orange, brown stripes)
- LED1: multicolor LED

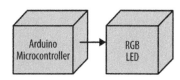

Figure 9-1. *The RGB Flasher block diagram*

Circuit Theory

Figure 9-2 shows a typical RGB LED with the wiring pinout names. There are three pins, one for each color, and one common pin for positive attachment to a power supply. Like the ordinary LED, the positive and negative pins are wired to the positive and negative points of a DC (direct current) circuit. To illustrate, Figure 9-3 shows three SPST (single pole, single throw) switches wired to control red, green, and blue LEDs. Closing the contacts on SPST switch SW1 will allow the battery's (VBattery) current to flow through the red LED, turning it on. The other switches (SW2 and SW3) will turn on the green and blue LEDs as well. The individual colors can be lit sequentially or at random using the three SPST switches. The Arduino microcontroller will provide a sequential switching order, allowing the red, green, and blue LEDs to turn on accordingly.

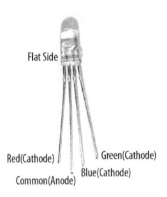

Figure 9-2. *A typical RGB LED with pinout names*

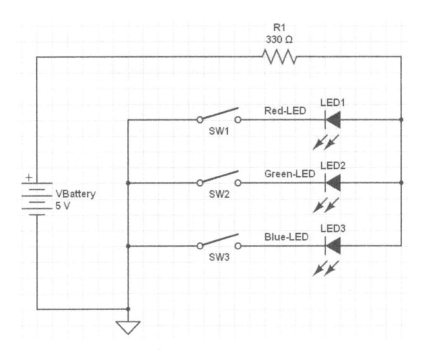

Figure 9-3. *Three SPST switches controlling an RGB LED*

Tech Note

A common anode RGB LED has all of the positive leads connected together to one lead.

The RGB Flasher

The RGB Flasher is an awesome Arduino microcontroller gadget that displays three colors (red, green, and blue) on one LED. You can easily build the circuit on the MakerShield, which will make it portable so you can carry it in your shirt pocket. You can use either the Fritzing diagram shown in Figure 9-4 or the circuit schematic diagram of Figure 9-5 to build the flasher.

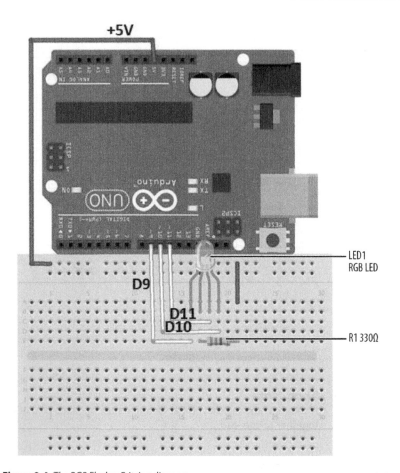

Figure 9-4. *The RGB Flasher Fritzing diagram*

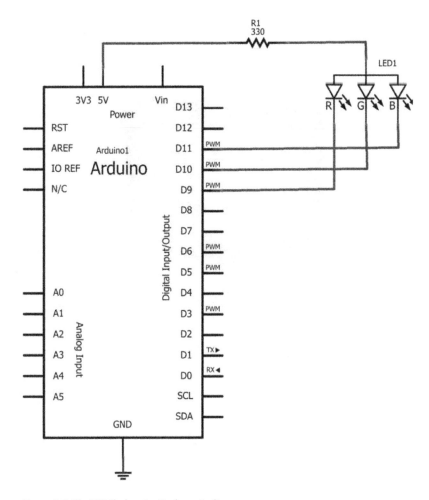

Figure 9-5. *The RGB Flasher circuit schematic diagram*

Tech Note

The common anode pin is the longest lead on the RGB LED.

After wiring the components onto the mini breadboard of the MakerShield pictured in Figure 9-6, upload Example 9-1 to the Arduino microcontroller.

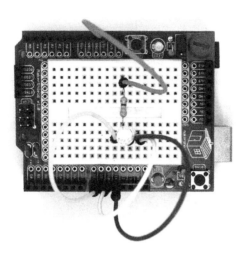

Figure 9-6. *The RGB Flasher built on a MakerShield*

Example 9-1. *The RGB Flasher sketch*

```
/*
  RGB Flasher

  Flashes the red, green, and blue LEDs of an RGB LED
  Turns on an LED on for one second, then off for one second for each
  color LED.

  15 Feb 2013
  Don Wilcher

 */

// RGB pins wired to the Arduino microcontroller.
// give them names:
int redled = 9;
int grnled = 10;
int bluled = 11;

// the setup routine runs once when you press reset:
void setup() {
  // initialize the digital pins as outputs:
  pinMode(redled, OUTPUT);
  pinMode(grnled, OUTPUT);
```

```
  pinMode(bluled, OUTPUT);
  // turn RGB outputs off:
  digitalWrite(redled, HIGH);
  digitalWrite(grnled, HIGH);
  digitalWrite(bluled, HIGH);
}

// the loop routine runs over and over again forever:
void loop() {

  digitalWrite(redled, LOW);   //  turn the red LED on
  delay(1000);                 //  wait for a second
  digitalWrite(redled, HIGH);  //  turn the LED off
  delay(1000);                 //  wait for a second
  digitalWrite(grnled, LOW);   //  turn the green LED on
  delay(1000);                 //  wait for a second
  digitalWrite(grnled, HIGH);  //  turn the green LED off
  delay(1000);                 //  wait for a second
  digitalWrite(bluled, LOW);   //  turn the blue LED on
  delay(1000);                 //  wait for a second
  digitalWrite(bluled, HIGH);  //  turn the blue LED off
  delay(1000);                 //  wait for a second
}
```

So far in this book, you've been taking an LED pin HIGH to light it. This example takes a pin LOW to light it. This is because the common pin on the RGB LED goes to +5V and each element's pin (R, G, and B) is a negative lead. As a result, each of those pins needs to be taken LOW to allow current to flow. This means that taking a pin HIGH turns it *off*. This is the reverse of what you've seen with discrete LEDs in this book where there is one positive and one negative lead. In the case of this RGB LED, there is one positive lead (the *common anode*) and three negative leads (*cathodes*).

After uploading the RGB Flasher sketch to the Arduino Microcontroller, the red, green, and blue LEDs will be individually flashing in sequence. You can change the order of the LEDs by making new sketch RGB pin assignments along with appropriate breadboard wiring changes. Like a good scientist, remember to record your observations, modified sketches, and circuit schematic diagrams in your lab notebook!

Troubleshooting Tip

If the LEDs don't turn on in the proper sequence, check your sketch pin assignments, as well as the orientation of the component on the MakerShield mini breadboard.

Something to Think About

Are there common cathode RGB LEDs? If so, what Arduino microcontroller wiring changes are needed to operate them correctly?

The Magic Light Bulb

10

Pushbutton Multicolor Flasher

Here's a cool trick you can play on a friend using the Arduino and an RGB LED. Build an RGB flasher with a mini pushbutton switch on an Arduino MakerShield. Show your friend the MakerShield and tell him the mini light bulb is magical and it can produce three colors: red, green, and blue. Have your friend close his eyes and chant "Are Gee Bee" three times. Briefly press the mini pushbutton switch to start the RGB flashing sequence. Tell your friend to open his eyes and watch in amazement the mini color light show produced by the Arduino. The Magic Light Bulb device is shown in Figure 10-1. The parts you will use for this project are two fixed resistors, an RGB LED, a pushbutton switch, and an Arduino microcontroller. The Magic Light Bulb will be built using the handy MakerShield. The Ultimate Microcontroller Pack has all of the parts for the project.

Parts List

- Arduino microcontroller
- MakerShield kit
- R1: 330Ω resistor (orange, orange, brown stripes)
- R2: 1KΩ resistor (brown, black, red stripes)
- LED1: RGB LED

Figure 10-1. *The Magic Light Bulb*

Let's Build a Magic Light Bulb

The Magic Light Bulb is an easy-to-build Arduino project using electronic parts from the Ultimate Microcontroller Pack. You can build the electronic circuit on a breadboard or the MakerShield. Building the Magic Light Bulb on the MakerShield allows the project to fit nicely in the palm of your hand, which offers great visual appeal for presentation to your friends. The wiring for the Magic Light Bulb can be constructed by using the Fritzing diagram shown in Figure 10-2.

Although the Fritzing diagram shows the breadboard and circuit components wired separately from the Arduino microcontroller, the project can easily be built on a MakerShield, as shown in Figure 10-1.

 Tech Note
Check your wiring for errors using the Fritzing diagram before applying power to the circuit.

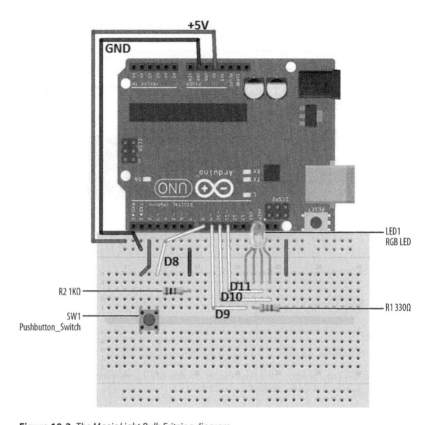

Figure 10-2. *The Magic Light Bulb Fritzing diagram*

Upload the Magic Light Bulb Sketch

With the Magic Light Bulb circuit built on the MakerShield, it's time to upload the sketch. Example 10-1 operates the RGB LEDs using a mini pushbutton switch. Here are the steps you'll need to follow:

1. Attach the Arduino to your computer using a USB cable.

2. Open the Arduino software and type Example 10-1 into the software's text editor.

3. Upload the sketch to the Arduino.

4. Press the mini pushbutton switch for a moment.

The Arduino will sequence the RGB LED tricolor pattern three times. Figure 10-3 shows the Magic Light Bulb in action.

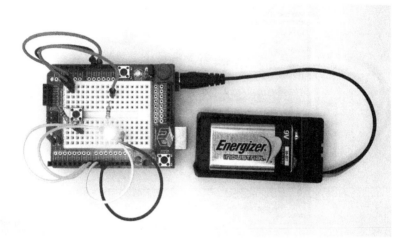

Figure 10-3. *The Magic Light Bulb running through the tricolor pattern*

Example 10-1. *The Magic Light Bulb sketch*

```
/*

  Magic Light Bulb

  Flashes the red, green, and blue LEDs of an RGB LED three times by
  briefly pressing a mini pushbutton switch.

  25 Feb 2013
  Don Wilcher

 */

// Pushbutton switch and RGB pins wired to the Arduino microcontroller.
// give them names:
int redled = 9;
int grnled = 10;
int bluled = 11;
int Pbutton = 8;
// initialize counter variable
 int n =0;
// monitor pushbutton switch status:
int Pbuttonstatus = 0;

// the setup routine runs once when you press reset:
void setup() {
// initialize the digital pins as outputs:
  pinMode(redled, OUTPUT);
  pinMode(grnled, OUTPUT);
  pinMode(bluled, OUTPUT);
// initialize the digital pin as an input:
```

```
   pinMode(Pbutton, INPUT);

// turn RGB outputs off:
   digitalWrite(redled, HIGH);
   digitalWrite(grnled, HIGH);
   digitalWrite(bluled, HIGH);
}
// the loop routine runs 3x after the pushbutton is pressed:
void loop() {
   Pbuttonstatus = digitalRead(Pbutton); //  read pushbutton status
   if(Pbuttonstatus == HIGH) {            //  if it's HIGH, start RGB Flasher
      for (n = 0; n < 3; n++){            //  flash RGB LEDs 3x
      digitalWrite(redled, LOW);          //  turn the red LED on (LOW is on)
      delay(250);                         //  wait for a 1/4 second
      digitalWrite(redled, HIGH);         //  turn the LED off (HIGH is off)
      delay(250);                         //  wait for a 1/4 second
      digitalWrite(grnled, LOW);          //  turn the green LED on
      delay(250);                         //  wait for a 1/4 second
      digitalWrite(grnled, HIGH);         //  turn the green LED off
      delay(250);                         //  wait for a 1/4 second
      digitalWrite(bluled, LOW);          //  turn the blue LED on
      delay(250);                         //  wait for a 1/4 second
      digitalWrite(bluled, HIGH);         //  turn the blue LED off
      delay(250);                         //  wait for a 1/4 second
      }
}
 else{                          // if pushbutton is LOW, turn LEDs off
      digitalWrite(redled, HIGH);
      digitalWrite(grnled, HIGH);
      digitalWrite(bluled, HIGH);
   }
}
```

Troubleshooting Tip

If the LEDs don't turn on in the proper sequence, check your sketch pin assignments as well as the orientation of the component on the MakerShield mini breadboard.

Circuit Theory

The Arduino controls the tricolor pattern that is sent to the RGB LED, as shown in Example 10-1. The RGB LED sequence or pattern is started with a short press of the mini pushbutton switch. Within the sketch is a counter that plays the pattern for a set number of times (in this case, three times). The replay number can easily be changed by replacing the counter's value of 3 to a different number.

The block diagram in Figure 10-4 shows the building blocks and the electrical signal flow for the Magic Light Bulb. Circuit schematic diagrams are used by electrical engineers to quickly build cool electronic devices. The equivalent circuit schematic diagram for the Magic Light Bulb is shown in Figure 10-5.

The for Loop and Counters

When you briefly press the mini pushbutton switch, the RGB cycles its color sequence three times before turning off. A "for" loop is behind the magic. Here's the Arduino sketch code:

```
for (n = 0; n < 3; n++)
```

How does it work? The counter starts at zero (n = 0) and checks to see if the count value is less than 3 (n < 3). If the counter is less than 3, the counter adds one to itself (n++) to get the next count value. When the counter reaches a value of 3, the Arduino turns off the RGB LED.

Change the "3" to a "4" in the preceding code, and observe the RGB LED after uploading the modified sketch to the Arduino. Remember to record your observations and modified sketches into your lab notebook!

Figure 10-4. *The Magic Light Bulb block diagram*

Something to Think About

What happens to the Magic Light Bulb if the mini pushbutton switch is pressed continuously?

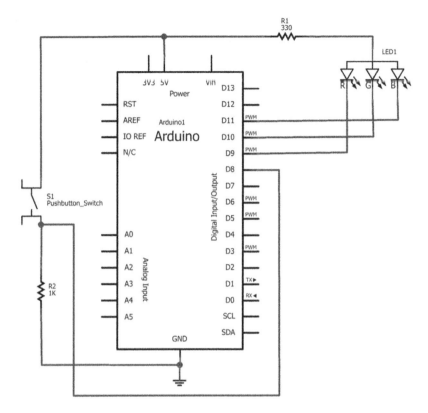

Figure 10-5. *The Magic Light Bulb circuit schematic diagram*

Metal Checker: The Electronic Switch

11

An Electronic Tester

Have you noticed household objects that appear to be made of metal when they're not? You and a friend can build an instrument that checks the metal properties of household objects in your home using an Arduino and a few electronic parts. The Arduino Metal Checker emits a tone when a metal object is placed across its test probes. With this awesome electronic instrument, you and a friend can uncover the metal mysteries hidden inside of your home. The Metal Checker uses an Arduino, two fixed resistors, one transistor, and a piezo buzzer. The Ultimate Microcontroller Pack makes it convenient to build the instrument because of the variety of electronic parts. The Metal Checker is shown in Figure 11-1.

Parts List

- Arduino microcontroller
- MakerShield kit
- Q1: 2N3904 NPN transistor
- R1: 330Ω resistor (orange, orange, brown stripes)
- R2: 1KΩ resistor (brown, black, red stripes)
- PB1: piezo buzzer

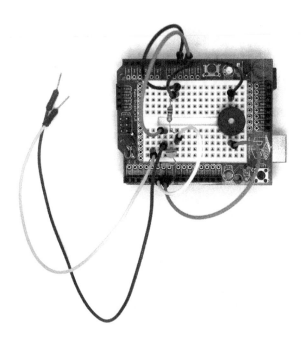

Figure 11-1. *The Metal Checker device*

Let's Build a Metal Checker

The Metal Checker is a cool electronics device to build with an Arduino and electronic parts from the Ultimate Microcontroller Pack. You can build the electronic circuit on an ordinary breadboard or the MakerShield. Building the Metal Checker on the MakerShield allows the device to fit nicely inside a Maker's toolbox or workbench drawers. Also, the MakerShield is small enough to carry with you in the field for scientific metal checking activities. Figure 11-2 provides a Fritzing diagram for building the Metal Checker.

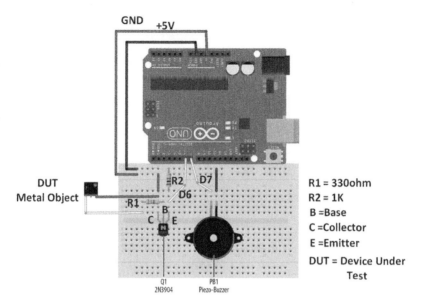

GND +5V

DUT
Metal Object

R2
D7
D6
R1
B
C E

R1 = 330ohm
R2 = 1K
B =Base
C =Collector
E =Emitter
DUT = Device Under
Test

Q1
2N3904

PB1
Piezo-Buzzer

Figure 11-2. *The Metal Checker Fritzing diagram*

Why Use a Transistor and Arduino for a Metal Checker?

A simple Metal Checker can easily be built using an LED, a battery, and wire. So why bother using an Arduino and a transistor? Based on the metal's electrical conductive properties, the transistor's external base resistor will set an appropriate sensing current to turn the transistor on. The transistor provides an approximate voltage value of +5VDC to the Arduino. Upon detecting the +5VDC signal, the Arduino turns on the piezo buzzer. Therefore, the transistor acts as an electronic switch, sensitive to certain amounts of electrical current flowing through the metal. The electronic switching and sensing functions can be adjusted based on the type of metal. Also, different piezo buzzer sounds can be programmed into the Arduino to reflect various metals as well.

A simple Metal Checker cannot be modified to have such cool detecting features because of the limited parts used.

The Metal Checker uses a small transistor for metal sensing. To correctly wire it with the Arduino, Figure 11-3 shows a picture with the proper transistor pinout. Use both the Fritzing diagram and the transistor pinout to ensure correct attachment to the Arduino.

E=Emitter   C = Collector

B = Base

Figure 11-3. *The 2N3904 NPN transistor pinout*

Although the Fritzing diagram shows the breadboard and electronic components wired separately from the Arduino, the device can easily be built on a MakerShield, as shown in Figure 11-1.

Tech Note
Check your wiring for errors using the Fritzing diagram before applying power to the circuit.

Upload the Metal Checker Sketch

With the Metal Checker circuit built on the MakerShield, it's time to upload the sketch. Example 11-1 operates the piezo buzzer using a small transistor. Here are the steps you'll need to follow:

1. Attach the Arduino to your computer using a USB cable.
2. Open the Arduino software and type Example 11-1 into the software's text editor.
3. Upload the sketch to the Arduino.
4. Touch the test probes together.

The Arduino will turn on the piezo buzzer. Now you're ready to unlock the metal mysteries hiding in your house!

Example 11-1. *The Metal Checker sketch*

```
/*
  Metal Checker

  Turns on and off a piezo buzzer at pin 7 when metal is placed across
  the sense wires of the metal sensor circuit attached to pin 6.

  The circuit:
    * Piezo buzzer attached from pin 7 to ground
    * Metal Checker sensor attached to pin 7
    * 1KΩ fixed resistor attached from pin 6 to ground

  March 2013
  by Don Wilcher


*/

// set pin numbers:
const int MSensePin = 6;     // the number of the metal sense pin
const int PBuzzerPin =  7;   // the number of the piezo buzzer pin

// variables will change:
int MetalStatus = 0;         // variable for the metal sense status

void setup() {
  // initialize the LED pin as an output:
  pinMode(PBuzzerPin, OUTPUT);
  // initialize the pushbutton pin as an input:
  pinMode(MSensePin, INPUT);
}

void loop(){
  // read the state of the metal sense value:
  MetalStatus = digitalRead(MSensePin);

  // check if metal is present
  // if it is, the MetalStatus is HIGH:
  if (MetalStatus == HIGH) {
    // turn piezo buzzer on:
    digitalWrite(PBuzzerPin, HIGH);
  }
  else {
    // turn MetalStatus off:
    digitalWrite(PBuzzerPin, LOW);
  }
}
```

The Transistor

The transistor is a small electronic component that can be used as an electronic switch or an amplifier. There are two types of transistors: NPN and PNP. These transistor types can be used as electronic switches and amplifiers: the current flow in the PNP transistor is opposite of that in the NPN transistor. The 2N3904 NPN transistor is being used as an electronic switch, replacing the mini pushbutton component of previous Arduino projects. To learn more about the transistor, read Charles Platt's *Make: Electronics* (Maker Media, 2009).

Troubleshooting Tip
If the piezo buzzer doesn't turn on, check your sketch pin assignments as well as the orientation of the transistor on the MakerShield mini breadboard.

Circuit Theory

The 2N3904 NPN transistor provides a signal to the Arduino, allowing it to turn on the piezo buzzer. Placing a metal object on the test probes allows electrical current to flow through the transistor, turning it on like a pushbutton switch. The 1KΩ (kilo-ohm) resistor provides a control voltage (+5VDC) to the Arduino, allowing it to turn on the piezo buzzer. Placing plastic objects on the test probes will not activate the transistor, the Arduino, and the piezo buzzer.

The block diagram in Figure 11-4 shows the building blocks and the electrical signal flow for the Metal Checker. A circuit schematic diagram used by electrical engineers to quickly build cool electronic devices is shown in Figure 11-5. Circuit schematic diagrams use electrical symbols for electronic components and are abbreviated drawings of Fritzing diagrams.

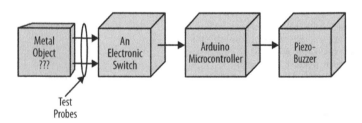

Figure 11-4. *The Metal Checker block diagram*

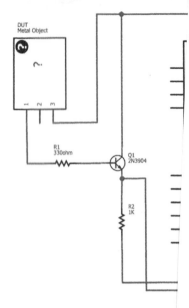

Figure 11-5. *The Metal Checker circuit*

Something to TI

Can a Metal Checker be used as a

Electrical Sa
Under no circum
powered electric
tection of the Ard
er to test any ele

The Theremin

12

A Simple Transistor Amplifier

Electronic circuits that produce audible sounds have been used to create strange and eerie audio effects for science-fiction movies like *Star Wars* and *Marvel's The Avengers*. The Theremin is a device that generates different electronic sounds by waving hands over and around a pair of protruding antennas.

You can make your own awesome Theremin using a few electronic components from the Ultimate Microcontroller Pack. The Theremin in this project will not use a pair of antennas but a photocell to interact with the device. Also, a simple transistor amplifier will be built to enhance the volume of the electronic sounds created with your Theremin. The Parts List shows all of the electronic components available from the Ultimate Microcontroller Pack to build your own Theremin. The Theremin built on a MakerShield is shown in Figure 12-1.

Parts List

- Arduino microcontroller
- MakerShield kit
- Q1: 2N3904 or S9013 NPN transistor
- R1: 10KΩ resistor (brown, black, orange stripes)
- R2: 1KΩ resistor (brown, black, red stripes)
- SPKR1: mini 8Ω speaker
- C1: 100 uF electrolytic capacitor
- PC1: photocell

Figure 12-1. *The Theremin*

Let's Build a Theremin

The Theremin, invented in 1920 by Russian inventor Leon Theremin, uses an electronic circuit called an oscillator to create different sounds. In our Theremin, we're using the Arduino as an oscillator by programming it to select different tones based on changing light levels. The tone changes are made by waving your hand over a photocell, creating various sounds based on changing light levels. The circuit is built on a breadboard with electronic components from the Ultimate Microcontroller Pack, as just shown in the Parts List. Although the Theremin can be built on an ordinary breadboard, the MakerShield makes the device small enough to carry in a shirt pocket or Maker bag. Figure 12-2 shows a Fritzing diagram of the Theremin. Also, the actual mini 8Ω speaker used in the Theremin project is shown in Figure 12-3.

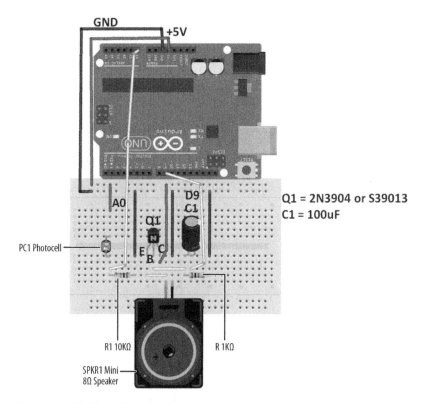

GND
+5V

A0
D9
C1

Q1
N
E   C
B

PC1 Photocell

Q1 = 2N3904 or S39013
C1 = 100uF

R1 10KΩ
R 1KΩ

SPKR1 Mini
8Ω Speaker

Figure 12-2. *The Theremin Fritzing diagram*

The electronic sounds generated by the Arduino are wired to a simple transistor amplifier. Pay close attention to the 100 uF electrolytic capacitor's orientation (shown on the Fritzing diagram) to prevent damage to the Arduino. Also, the NPN transistor's pinout for either a 2N3904 or S9013 electronic component is shown on the Fritzing diagram's breadboard. The mini 8Ω speaker color wire leads must be connected correctly (as shown in Figure 12-2) in order for the audio electronic sounds to be heard through it.

Figure 12-3. *The mini 8Ω speaker*

Tech Note
The 100 uF electrolytic capacitor is called a polarized capacitor because of its positive and negative electrical leads. Like an LED, electricity flows through it easily in only one direction.

Upload the Theremin Sketch

It's time to upload the sketch to the Arduino with the Theremin's photocell and simple transistor amplifier circuits built on the MakerShield. Example 12-1 operates the Arduino-based Theremin using a photocell and a simple transistor amplifier circuit. Here are the steps you'll need to follow:

1. Attach the Arduino to your computer using a USB cable.
2. Open the Arduino software and type Example 12-1 into the software's text editor.
3. Upload the sketch to the Arduino.

Once the Theremin sketch has been uploaded to the Arduino, the mini 8Ω speaker will begin emitting an electronic buzzing sound. A wave of your hand over the photocell will change the tone coming from the speaker. With a small amount of light on the photocell, the tone's pitch will decrease. Placing the Theremin under a light bulb increases the electronic sound's pitch. Like the inventor Leon Theremin, now you're ready to create some really cool sounds from your Arduino-powered Theremin. Remember to document your new designs and experiments in a lab notebook!

Example 12-1. *The Theremin sketch*

```
/*
  Theremin

  Plays sound effects through a simple transistor
  amplifier using a photocell.

  I/O circuits:
   * A simple transistor amplifer wired to digital pin 8
   * A photocell wired to analog 0 and +5V
   * A 10K resistor wired to analog 0 pin to ground

*/


void setup() {
  // initialize serial communications (for debugging only):
  Serial.begin(9600);
}

void loop() {
  // read the sensor:
  int sensorReading = analogRead(A0);
  // print the sensor reading so you know its range:
  Serial.println(sensorReading);

  // map the analog input range (in this case, 400-1000 from
  // the photoresistor) to the output pitch range (120-1500 Hz)
  // change the minimum and maximum input numbers below
  // depending on the range your sensor's giving:
  int thisPitch = map(sensorReading, 400, 1000, 120, 1500);

  // play the pitch:
  tone(9, thisPitch, 10);
  delay(1);        // delay in between reads for stability
}
```

Troubleshooting Tip

If the mini 8Ω speaker doesn't emit sound, check your sketch for programming errors and also check the orientation of the transistor and the electrolytic capacitor on the MakerShield mini breadboard.

The Serial Monitor

Sometimes you may have to debug or troubleshoot a sketch because of programming errors. The Serial Monitor is an embedded tool of the Arduino software that allows you to display information from your program, such as data variables, on your computer screen. To display the data, use the simple Arduino instruction `Serial.println()`. The Theremin sketch uses this to display the photocell sensor data. As shown in Figure 12-4, the sketch instruction provides scrolling sensor data from the photocell to the Serial Monitor.

To access the data monitor, add the `Serial.begin(9600)` instruction within the `void setup()` function to open a communication link between your computer and the Arduino. The `9600` sets the speed at which the Arduino sends data to your computer. It means your Arduino is sending data to your computer at 9,600 bits every second. (Another term used to describe the data transmission speed is "Baud" rate.)

To display variables on the Serial Monitor, use the instruction `Serial.print In(variable name)`.

Figure 12-4. *Photocell sensor data scrolling on the Serial Monitor*

Circuit Theory

The 2N3904 or S39013 NPN transistor amplifies or increases the audio signal created by the Arduino. The transistor has an amplification value called "gain" used to determine the volume of an electrical signal. A typical gain value engineers use in designing simple amplifiers like this one is 100. The mini 8Ω speaker can be wired directly to pin D9 with a reasonable amount of volume, but the simple transistor amplifier increases the sound by a factor of 100, making the Theremin sound louder.

The block diagram in Figure 12-5 shows the building blocks and the electrical signal flow for the Theremin. A Fritzing software circuit schematic diagram of the Theremin is shown in Figure 12-6. As a reminder, circuit schematic diagrams use electrical symbols for electronic components and are abbreviated drawings of Fritzing diagrams.

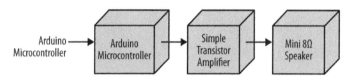

Figure 12-5. *The Theremin block diagram*

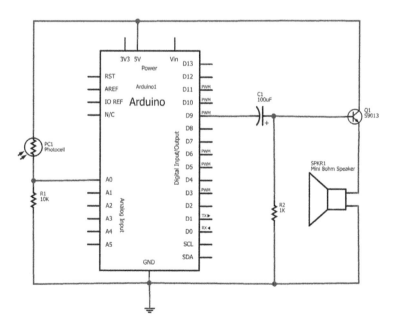

Figure 12-6. *The Theremin circuit schematic diagram*

Something to Think About

What sounds would be emitted by the Theremin's simple transistor amplifier if the mini 8Ω speaker was replaced with a piezo buzzer? Try it!

An Arduino Ohmmeter 13

The Ultimate Microcontroller Pack has a supply of resistors you can use in your projects. These resistors are color coded to indicate their resistive value. If you already know how to read the color code—or once you learn how—you'll be able to glance at a resistor and tell automatically what its value is. But what about other components in your projects? What about the LEDs, potentiometers, buzzers, or even the wires themselves? How much resistance do they add to the circuit? To find the answer, you'll need an ohmmeter, a device that measures the resistance of an electrical component. With a resistor, a breadboard, and a few wires, you can turn your Arduino into this useful, awesome measuring device. The Arduino Ohmmeter is shown in Figure 13-1.

Parts List

- Arduino microcontroller
- MakerShield kit
- R1: 1KΩ resistor (brown, black, orange stripes)
- R2: other resistors chosen at random

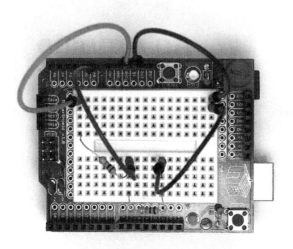

Figure 13-1. *An Arduino Ohmmeter*

Let's Build an Arduino Ohmmeter

This gadget tests the resistance of electrical components. Place the unknown resistor you want to test in series with the reference resistor R1 connected to GND. The Arduino will calculate the resistance and display it on the Serial Monitor. The resistance of other electrical objects can be measured with the Arduino Ohmmeter as well. Building the Arduino Ohmmeter on a MakerShield protoboard makes the device small enough to carry to a friend's house to check his electronic projects. Figure 13-2 shows the Fritzing diagram for the Arduino Ohmmeter.

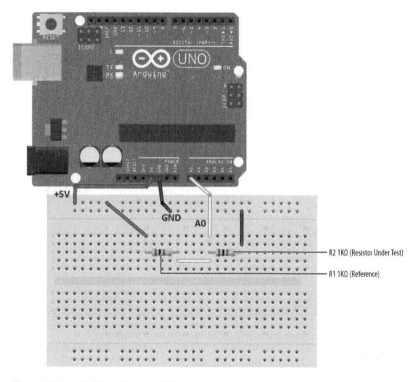

Figure 13-2. *An Arduino Ohmmeter Fritzing diagram*

Upload the Arduino Ohmmeter Sketch

It's time to upload the Ohmmeter sketch to the Arduino. Example 13-1 reads the resistance of R2, and reports the result through the serial display. Here are the steps you'll need to take:

1. Attach the Arduino to your computer using a USB cable.

2. Open the Arduino software and type Example 13-1 into the software's text editor.

3. Upload the sketch to the Arduino.

Once the Ohmmeter sketch has been uploaded to the Arduino, place the unknown resistor (shown as R2 on the Frizting diagram) you want to test in series with the reference resistor R1 (1KΩ) connected to GND. The voltage across the R2 resistor and its resistance value will be displayed on the Serial Monitor. Figure 13-3 shows the output voltage (Vout) and the measured resistance of a 1KΩ resistor (R2) being displayed on the Serial Monitor.

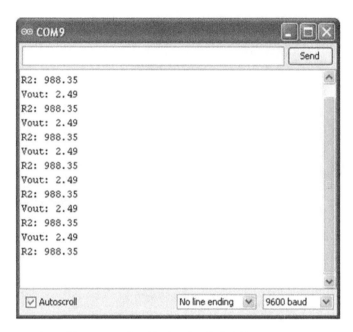

Figure 13-3. *R2 and Vout measured and displayed on the Serial Monitor*

Example 13-1. *The Arduino Ohmmeter sketch*

```
/*
  Arduino Ohmmeter

 */

// set up pins on Arduino for LED and test lead
int analogPin = 0;      // reads the resistance of R2
int raw = 0;            // variable to store the raw input value
int Vin = 5;            // variable to store the input voltage
float Vout = 0;         // variable to store the output voltage
float R1 = 1000;        // variable to store the R1 value
float R2 = 0;           // variable to store the R2 value
float buffer = 0;       // buffer variable for calculation

void setup()
{
  Serial.begin(9600);              // Set up serial

}

void loop()
{
  raw = analogRead(analogPin);  // reads the input pin
  if(raw)
    {
```

```
        buffer = raw * Vin;
        Vout = buffer /1024.0;  // calculates the voltage on the input pin
        buffer = (Vin / Vout) - 1;
        R2 = R1 / buffer;
        Serial.print("Vout: ");
        Serial.println(Vout);         // outputs the information
        Serial.print("R2: ");         //
        Serial.println(R2);           //
        delay(1000);
    }
}
```

Tech Note

The value of R1 is stored using a float variable type in the sketch. Change the value from 1000 (1KΩ) to a higher value when reading higher resistance values.

Circuit Theory

The operation of the Arduino Ohmmeter is based around the concept of the *voltage divider*. Two resistors are connected in series, and the reading is taken from where the two resistors join. The voltage measured at that point is the ratio of R2/(R1+R2) multiplied by the voltage in. For example, if R2 is 10K and R1 is 10K, then the ratio is 1/2; multiplying that by 5 volts returns 2.5 volts.

The Arduino Ohmmeter uses that relationship between voltage and resistance slightly differently. It knows that it started with 5 volts from the VCC pin. It also knows that the reference R1 has a value of 1K ohms. It then reads the divided voltage in at analog pin 0. Using those numbers, it is relatively easy to calculate the value of the object at R2.

A Fritzing electronic circuit schematic diagram of the Ohmmeter is shown in Figure 13-4. The Arduino Ohmmeter block diagram showing the connecting electronic components is shown in Figure 13-5. As a reminder, circuit schematic diagrams use electrical symbols for electronic components and are abbreviated drawings of Fritzing diagrams.

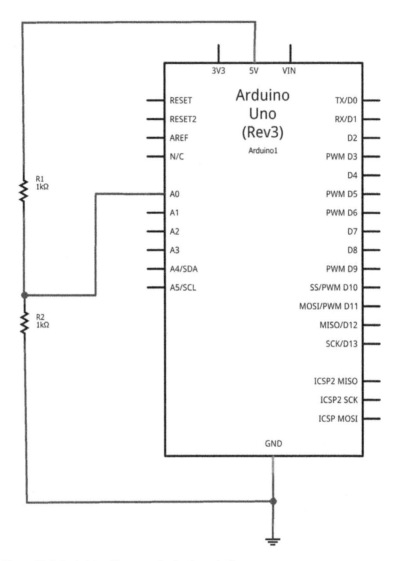

Figure 13-4. *An Arduino Ohmmeter circuit schematic diagram*

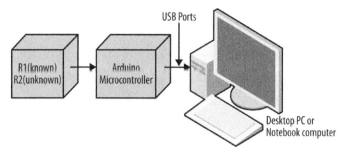

Figure 13-5. *An Arduino Ohmmeter circuit block diagram*

Electrical Safety Tip

Under no circumstance should the Ohmmeter be used to test powered electrical/electronic devices. For your safety and the protection of the Arduino and MakerShield, do not use the Ohmmeter to test any powered electrical/electronic devices!

Something to Think About

How can a small slide switch be added to select between two unknown resistors for measuring their resistance?

The LCD News Reader 14.

Have you ever wondered what cool projects you can build with the Ultimate Microcontroller Pack LCD (liquid crystal display)? So far in this book, the Arduino has communicated with us via sound, via blinking LEDs, and via the Serial Monitor. What would it be like if the Arduino could communicate through a self-contained screen that could display two lines of text at a time? This project is all about using the LCD to display information in characters made of letters, numbers, and a few special symbols. You can make the information scroll and reverse, and you can even do some very simple animations! We start learning about the LCD in Figure 14-1.

Parts List

- Arduino microcontroller
- Full-size clear breadboard
- R1: 10KΩ potentiometer
- LCD1: LMB162ABC 16x2 LCD
- 16-pin male header (electrical connector)

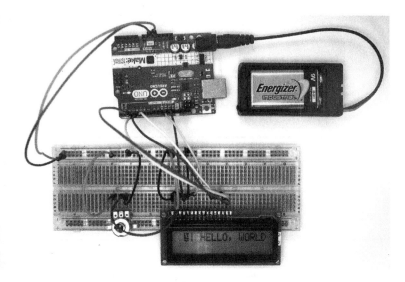

Figure 14-1. *The LCD News Reader*

Let's Build the LCD

The first task in building the LCD News Reader is to solder a 16-pin male header to the LCD. The Ultimate Microcontroller Pack has several male headers for building your own Arduino shields. The header needs to be cut to a length to match the 16 LCD copper pad holes. Figure 14-2 shows the male header cut to the appropriate LCD length. Insert the 16-pin male header through the copper pad holes and solder them one by one to the LCD printed circuit board (PCB). Figure 14-3 shows the male header soldered onto the LCD PCB.

Place the LCD onto the solderless breadboard, as shown in Figure 14-4. Wire LCD pin number "1" to ground and "2" to +5VDC. Attach the center pin of the 10KΩ potentiometer to pin number "3" of the LCD. Wire the remaining 10KΩ potentiometers pins to +5VDC and ground as shown in the diagram. With the LCD wired to the solderless breadboard, apply power to it using the Arduino. Adjust the 10KΩ potentiometer until the LCD's top row displays pixel squares, as shown in Figure 14-4. Complete the rest of the tester wiring using the Fritzing diagram shown in Figure 14-5.

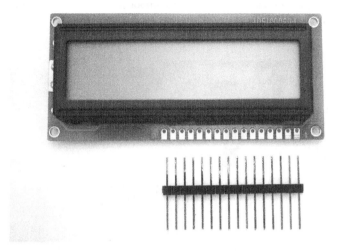

Figure 14-2. *The 16-pin male header cut to match the length of the LCD copper pad holes*

 Tech Note
A header is an electrical connector.

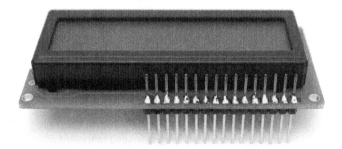

Figure 14-3. *Male header soldered onto the LCD PCB*

Figure 14-4. *Testing the LCD wiring on a full-size clear breadboard*

Upload the LCD News Reader Sketch

It's time to upload the sketch for the LCD News Reader to the Arduino. Here are the steps you'll need to follow:

1. Attach the Arduino to your computer using a USB cable.
2. Open the Arduino software and type Example 14-1 into the software's text editor.
3. Upload the sketch to the Arduino.

Once the LCD News Reader sketch has been uploaded to the Arduino, the LCD will display a message, as shown in Figure 14-1. According to computing tradition, the first message you should display on a new piece of hardware is "Hello, World!" Figure 14-6 shows the LCD News Reader displaying various screens.

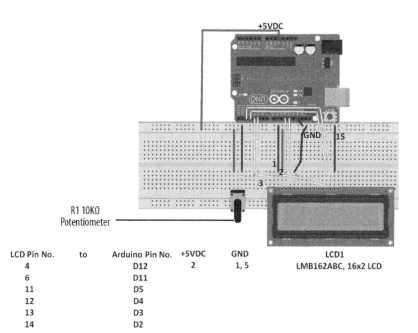

LCD Pin No.	to	Arduino Pin No.	+5VDC	GND	LCD1
4		D12	2	1, 5	LMB162ABC, 16x2 LCD
6		D11			
11		D5			
12		D4			
13		D3			
14		D2			

Figure 14-5. *The LCD News Reader Fritzing diagram*

Example 14-1. *The LCD News Reader sketch*

```
/*
  The LCD News Reader

  20 August 2013

*/

// include the LCD library code:
#include <LiquidCrystal.h>

// set up pins on Arduino for LCD and test lead
LiquidCrystal lcd(12,11,5,4,3,2);


// set up the LCD's number of columns and rows

#define Xdelay 1900

String a;
String b;
String c;
String d;
```

```
void setup() {
  lcd.begin(16,2);
  lcd.setCursor(0,0);


  clearLCD();
  backlightOn();

  lcd.print("HELLO, WORLD!");
  delay(Xdelay);

}


void loop()
{

  char databuff[16];
  char dispbuff[16];

  // display on/off test
  for(int x = 5; x>0; x--)
      {
      delay(1000);
      displayOff();
      delay(1000);
      displayOn();
      }


  clearLCD();
  backlightOn();
  lcd.print("SLOW FADE        ");
  fadeOut(100);
  fadeIn(10);

  // light up all segments as a test



  lcd.print("0123456789abcdef");
  delay(Xdelay);
  lcd.print("ghijklmnopqrstuv");
  delay(Xdelay);
  lcd.print("wxyz +?*&%$#()!=");
  delay(Xdelay);
  lcd.print("                ");
  delay(Xdelay);
  lcd.print("                ");
  delay(Xdelay);

  a = "0123456789abcdef";
  b = "ghijklmnopqrstuv";
  c = "wxyz +?*&%$#()!=";
  d = "                ";
```

```
selectLineTwo();
lcd.print(a);
delay(Xdelay);

selectLineOne();
lcd.print(a);
selectLineTwo();
lcd.print(b);
delay(Xdelay);

selectLineOne();
lcd.print(b);
selectLineTwo();
lcd.print(c);
delay(Xdelay);

selectLineOne();
lcd.print(c);
selectLineTwo();
lcd.print(d);
delay(Xdelay);

selectLineOne();
lcd.print(d);
selectLineTwo();
lcd.print(d);
delay(Xdelay);


for (int x = 0; x<=5; x++)
  {
  for(int i = 15; i>=0; i--)
    {
      goTo(i);
      if (i%4 == 1)
        lcd.print("- ");
      if (i%4 == 2)
        lcd.print("I ");
      if (i%4 == 3)
        lcd.print("- ");
      if (i%4 == 0)
        lcd.print("I ");
      delay(100);
    }
  for(int i =0; i<=14; i++)
    {
      goTo(i);
      lcd.print(" @");
      delay(100);
    }
  }

clearLCD();
}
```

```
void selectLineOne()
{
   lcd.write(0xFE);   //command flag
   lcd.write(128);    //position
   delay(10);
}
void selectLineTwo()
{
   lcd.write(0xFE);   //command flag
   lcd.write(192);    //position
   delay(10);
}
void goTo(int position)
{
if (position<16)
  {
    lcd.write(0xFE);   //command flag
    lcd.write((position+128));     //position
  }else if (position<32)
    {
     lcd.write(0xFE);   //command flag
     lcd.write((position+48+128));     //position
} else { goTo(0); }
   delay(10);
}

void clearLCD()
{
   lcd.write(0xFE);   //command flag
   lcd.write(0x01);   //clear command
   delay(10);
}
void backlightOn()
{
    lcd.write(0x7C);   //command flag for backlight stuff
    lcd.write(157);    //light level
   delay(10);
}
void backlightOff()
{
    lcd.write(0x7C);   //command flag for backlight stuff
    lcd.write(128);    //light level for off
   delay(10);
}

void backlightValue(int bv)
{
    int val = bv;
    if (bv < 128) val= map(bv, 0, 1023, 128, 157);
    if (bv > 157) val = map(bv, 0, 1023, 128, 157);

    lcd.write(0x7C);   //command flag for backlight stuff
    lcd.write(val);    //light level
   delay(10);
```

```
}

void displayOn()
{
   lcd.write(0xFE);    //command flag
   lcd.write(0x0C);    //clear command
   delay(10);
}

void displayOff()
{
   lcd.write(0xFE);    //command flag
   lcd.write(0x08);    //clear command
   delay(10);
}

void fadeOut(int fade)
{
  for (int x = 157; x>128; x--)
  {
    backlightValue(x);
    delay(fade);
  }
}

void fadeIn(int fade)
{
  for (int x = 128; x<=157; x++)
  {
    backlightValue(x);
    delay(fade);
  }
}
```

Figure 14-6. *The LCD News Reader displaying various screens*

 Tech Note
A 16x2 LCD has two rows capable of displaying 16 characters each.

Circuit Theory

Example 14-1 displays a variety of characters, letters, and numbers based on C language programming instructions. The sketch is programmed to test all segments of the LCD as it cycles through the Arduino program. The Arduino sketch uses digital data pins D2, D3, D4, D5, D11, and D12 of its microcontroller chip to send text message information to the LCD. Time delays programmed into the sketch allow the characters, letters, and numbers to be displayed continuously on the LCD. The 10K potentiometer lets you adjust the contrast of the display.

The block diagram in Figure 14-7 shows the building blocks and the electrical signal flow for the LCD News Reader. Circuit schematic diagrams are used by electrical engineers to quickly build cool electronic devices. The equivalent circuit schematic diagram for the LCD News Reader is shown in Figure 14-8.

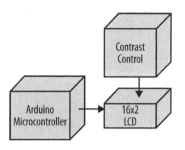

Figure 14-7. *The LCD News Reader block diagram*

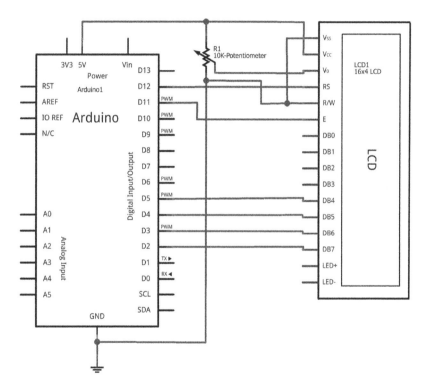

Figure 14-8. *The LCD News Reader circuit schematic diagram*

Something to Think About

How can a pushbutton switch be used to control the display?

A Logic Tester (with an RGB LED)

The NOT, AND, and OR projects (Chapters 5, 6, and 7, respectively) use two basic voltages: either +5VDC for TRUE or 0V for FALSE. These two signals let the Arduino make basic logic decisions. In the computer world, these voltages are known as binary data. In computers, binary data is represented by logic "1" (+5 volts DC) and logic "0" (0 volts). You can build a cool electronic device to see binary data using a few electronic components from the Ultimate Microcontroller Pack. The electronic components to build this device are shown in the Parts List. The Logic Tester with an RGB LED is shown in Figure 15-1.

Parts List

- Arduino microcontroller
- MakerShield kit
- R1: 1KΩ resistor (brown, black, red stripes)
- R2: 330Ω resistor (orange, orange, brown stripes)
- PB1: pushbutton switch
- LED1: RGB (red, green, blue) LED
- Long test wire

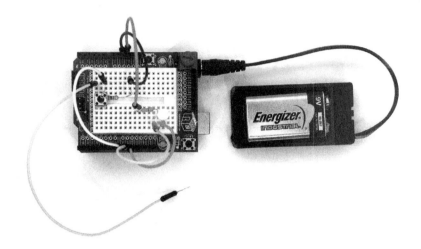

Figure 15-1. *A Logic Tester with an RGB LED*

Let's Build a Logic Tester

The Logic Tester is an easy-to-build Arduino microcontroller device. The RGB has three individual color LEDs that allow binary data to be seen visually. The RGB LED pinout is shown in Figure 15-2. Only the red and green LEDs will be used to show the binary logic values of "0" and "1". To ensure proper operation of the RGB LED, the flat side of the LED should be facing the Test pushbutton switch. See Figure 15-2 for the proper orientation of the RGB LED. Some LEDS may have the blue and green leads swapped. If yours is like that, you may need to move the G wire to the pin labeled B.

Tech Note
Base 2 is the number format for binary data.

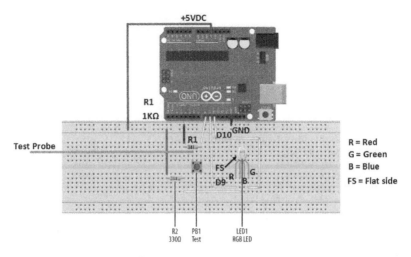

Figure 15-2. *Fritzing diagram for a logic tester with an RGB LED*

Upload the Logic Tester Sketch

With the Logic Tester built, it's time to upload the sketch. As shown in Example 15-1, the sketch operates an RGB LED using a pushbutton switch and two fixed resistors. Here are the steps you'll need to follow:

1. Attach the Arduino to your computer using a USB cable.

2. Open the Arduino software and type Example 15-1 into the software's text editor.

3. Upload the sketch to the Arduino.

Once the Logic Tester sketch has been uploaded to the Arduino microcontroller, the RGB's red LED will be on, as shown in Figure 15-1. Attaching the long test wire to the +5VDC source on the MakerShield and pressing the pushbutton switch will allow the RGB green LED to turn on, as shown in Figure 15-3.

Example 15-1. *The Logic Tester sketch*

```
/*
Logic Tester with RGB LED

Turns on the green LED when a logic "1" (+5V) signal is detected. The
red LED will turn on at logic "0" (0V) signal.  Also, when powering
up the Arduino the red LED is on.

4 May 2013
Don Wilcher

*/
```

```
// RG pins wired to the Arduino microcontroller
// give them names:
int redled = 9;
int grnled = 10;
int probein = 8;
int probeStatus = 0;

// the setup routine runs once when you press reset:
void setup() {
 // initialize the digital pins as outputs:
  pinMode(redled, OUTPUT);
  pinMode(grnled, OUTPUT);
  pinMode(probein, INPUT);
  // turn RGB outputs off:
  digitalWrite(redled, HIGH);
  digitalWrite(grnled, HIGH);

}

// the loop routine runs over and over again forever:
void loop() {
// read the status of the test probe value:
probeStatus = digitalRead(probein);

if (probeStatus == HIGH) { // check if the test probe value is HIGH

  digitalWrite(redled, HIGH);        // turn the red LED off (HIGH is off)
  digitalWrite(grnled, LOW);         // turn the green LED on (LOW is on)
 }
else {

   digitalWrite(redled, LOW);        // turn the red LED on
   digitalWrite(grnled, HIGH);       // turn the green LED off

 }
}
```

 Tech Note

HIGH is equivalent to binary 1 and LOW is equivalent to binary 0.

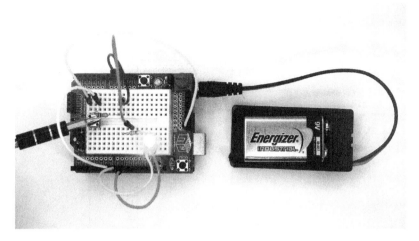

Figure 15-3. *The Logic Tester checking +5VDC on MakerShield*

Circuit Theory

Pressing the pushbutton will close the switch and allow +5 volts DC electrical current to flow through the test circuit. The Arduino reads digital pin 8 to determine if the pin is receiving +5 volts (i.e., set to HIGH) or if it is not receiving any voltage (i.e., set to LOW). The Arduino takes that information and lights up the appropriate LED: the green LED indicates that the pin is receiving +5 volts, and the red LED indicates 0 voltage. The digital pins used to operate the RGB's red and green LEDs are D9 and D10.

The block diagram in Figure 15-4 shows the electronic component blocks and the electrical signal flow for the Logic Tester. A Fritzing electronic circuit schematic diagram of the tester is shown in Figure 15-5. Electronic circuit schematic diagrams are used by electrical/electronic engineers to design and build cool electronic products for society.

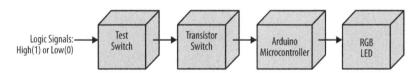

Figure 15-4. *The Logic Tester block diagram*

Tech Note

A block diagram is used to show electrical signal flow of electronic products.

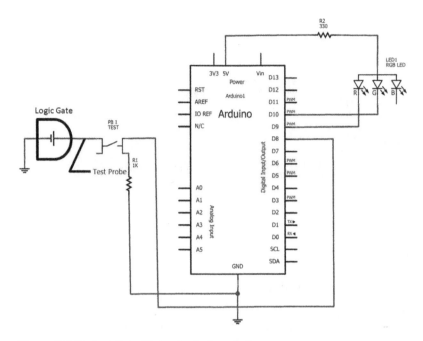

Figure 15-5. *The Logic Tester Fritzing circuit schematic diagram*

Something to Think About

How can the Logic Tester be operated without a pushbutton switch?

A Logic Tester (with an LCD) 16

The Logic Tester project in Chapter 15 allowed you to check the digital data values the Arduino uses to control motors and LEDs. The tester's two LEDs offer a quick way to see the digital data. In this project, you'll make an awesome change to the tester by displaying "HIGH (1)" or "LOW (0)" data messages on an LCD. The electronic components to build this device are shown in the Parts List. The Logic Tester with an LCD is shown in Figure 16-1.

Parts List

- Arduino microcontroller
- Full breadboard
- R1: 10KΩ potentiometer
- R2: 1KΩ resistor (brown, black, red stripes)
- R3: 330Ω resistor (orange, orange, brown stripes)
- Q1: S39014 NPN transistor or equivalent
- PB1: pushbutton switch
- One long jumper wire
- LCD1: LMB162ABC 16x2 LCD (liquid crystal display) with soldered 16-pin male header (electrical connector)

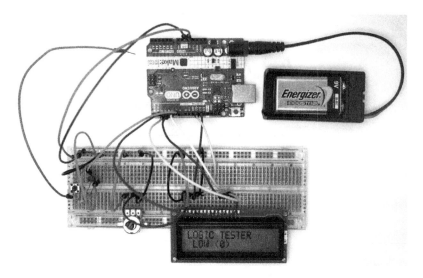

Figure 16-1. *A Logic Tester with an LCD*

Let's Build a Logic Tester

Building this tester requires the use of an LCD. If this is your first time using an LCD, I suggest reading Chapter 14. For help adding the 16-pin male header to the LCD, see Figure 14-2 and Figure 14-3. The 10KΩ potentiometer's center pin is wired to pin number 3 of the LCD. The potentiometer's remaining pins should be wired to +5VDC and ground. Place the LCD onto the solderless breadboard, as shown in Figure 16-2. LCD pin numbers 1 and 2 are wired to ground and +5VDC, respectively. Adjust the 10KΩ potentiometer contrast control for the LCD for proper pixel-square visibility. For reference on how to do this, see Figure 14-4.

Complete the rest of the tester wiring using the Fritzing diagram shown in Figure 16-2.

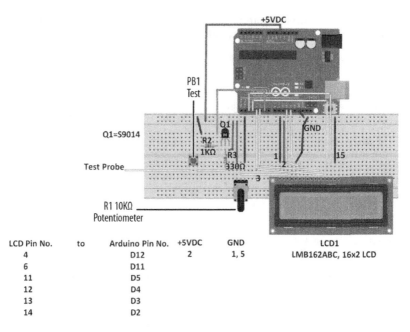

LCD Pin No.	to	Arduino Pin No.	+5VDC	GND	LCD1
4		D12	2	1, 5	LMB162ABC, 16x2 LCD
6		D11			
11		D5			
12		D4			
13		D3			
14		D2			

Figure 16-2. *Fritzing diagram for a Logic Tester with an LCD*

 Tech Note

Want to learn more about digital logic? Read Experiment 19 in Charles Platt's *Make: Electronics*.

Upload the Logic Tester Sketch

With the Logic Tester built, it's time to upload the sketch. Example 16-1 operates an LCD using a pushbutton switch, a transistor, and two fixed resistors. Here are the steps you'll need to follow:

1. Attach the Arduino to your computer using a USB cable.

2. Open the Arduino software and type Example 16-1 into the software's text editor.

3. Upload the sketch to the Arduino.

Once the Logic Tester sketch has been uploaded to the Arduino, the LCD will display a message, as shown in Figure 16-1. Take the long jumper wire (test probe) from the pushbutton switch and attach it to the +5V source of the Arduino (for reference, see Figure 16-2). Press the pushbutton switch and the LCD will display "HIGH (1)" for the

+5V source, as shown in Figure 16-3. Impress the local Makerspace by testing Arduino and digital electronic circuits with your Logic Tester!

Example 16-1. *The Logic Tester sketch*

```
/*
  Logic Tester
  LCD displays "HIGH (1)" when digital circuit signal is +5V. A "LOW (0)"
  is displayed when digital circuit signal is 0V.

  27 April 2013
  Don Wilcher

 */

// include the LCD library code:
#include <LiquidCrystal.h>

// set up pins on Arduino for LCD and transistor lead:
LiquidCrystal lcd(12,11,5,4,3,2);
int xistorPin = 6;
int digitalStatus = 0;       // variable for reading the digital circuit state


// initialize the transistor pin as an input and set up the LCD's number
// of columns and rows:
void setup() {
  lcd.begin(16,2);
  lcd.setCursor(0,0);
  lcd.print("LOGIC TESTER");
  pinMode(xistorPin, INPUT);

}


void loop() {
  // check if digital signal is HIGH or LOW:
digitalStatus = digitalRead(xistorPin);
if (digitalStatus == HIGH) {
  // if digital circuit signal is +5V, display HIGH (1):
  lcd.setCursor(0,1);
  lcd.print("HIGH (1) ");    // display HIGH (1)
}
else {
  // if digital circuit signal is 0V, display LOW (0):
  lcd.setCursor(0,1);
  lcd.print(" LOW (0) ");
 }
}
```

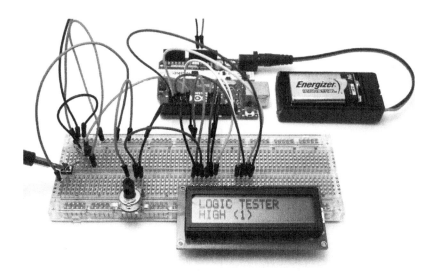

Figure 16-3. *The Logic Tester testing the Arduino's +5V source*

Tech Note

An electrical tester used to check digital circuits is called a logic probe.

Circuit Theory

Pressing the pushbutton will close the switch and allow +5 volts DC electrical current to flow through the circuit. The Arduino reads digital pin 6 to determine if the pin is receiving +5 volts (i.e., set to HIGH), or if it is not receiving any voltage (i.e., set to LOW). The Arduino takes that information and sends it to the LCD display via digital pins D2, D3, D4, D5, D11, and D12. The LCD then displays "HIGH (1)" or "LOW (0)" depending on the state of digital pin 6.

The block diagram in Figure 16-4 shows the electronic component blocks and the electrical signal flow for the Logic Tester. A Fritzing electronic circuit schematic diagram of the tester is shown in Figure 16-5. Electronic circuit schematic diagrams are used by electrical/electronic engineers to design and build cool electronic products for society.

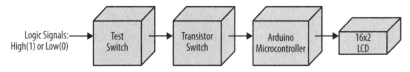

Figure 16-4. *The Logic Tester block diagram*

Tech Note

The initial LCD message displayed before testing a digital circuit is "LOW(0)."

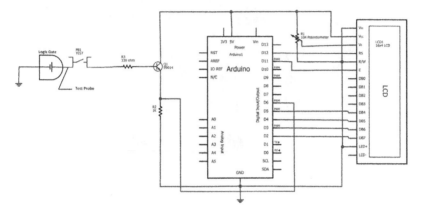

Figure 16-5. *The Logic Tester Fritzing circuit schematic diagram*

Something to Think About

How can a small piezo buzzer be used with the Logic Tester?

The Amazing Pushbutton (with Processing)

17

With this project, you can make colorful lines and numbers move up and down your computer screen as you press a simple pushbutton switch. To do that, this project will introduce you to Processing, a simple, easy-to-learn programming language (very much like the Arduino language) that makes it *very* easy to display graphics on a computer screen. When you connect an Arduino to Processing, you can make your Arduino draw on a screen!

The electronic components to build this device are shown in the Parts List. The Amazing Pushbutton is shown in Figure 17-1.

Parts List

- Arduino microcontroller
- MakerShield kit
- R1: 1KΩ resistor (brown, black, red stripes)
- PB1: pushbutton switch
- USB cable

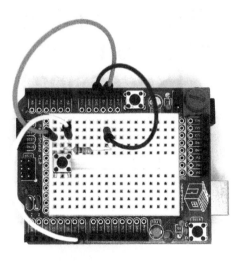

Figure 17-1. *The Amazing Pushbutton*

Let's Build an Amazing Pushbutton

Building the Amazing Pushbutton requires the use of a USB cable to send digital information from the Arduino to a computer screen. As shown in Figure 17-1, the device is quite simple to build, using only a 1KΩ fixed resistor and a pushbutton switch. The two components are connected in series. Where the two electronic components tie together, a jumper wire connects between them and pin D7 of the Arduino microcontroller.

Complete the rest of the Amazing Pushbutton wiring using the Fritzing diagram shown in Figure 17-2. The placement of the parts is not critical, so experiment with the locations of the electronic components and electrical wiring of the device. Although a mini breadboard is shown in the Fritzing diagram, the MakerShield protoboard provides a compact way to wire the device.

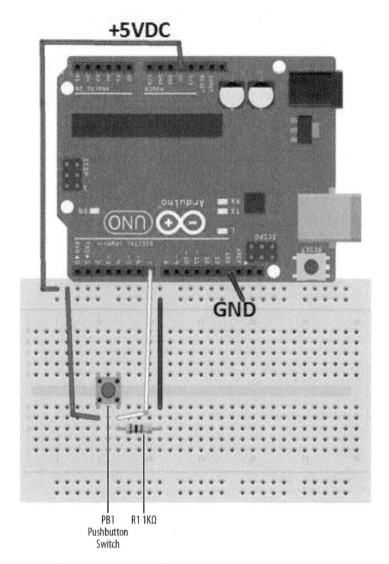

+5VDC

GND

PB1   R1 1KΩ
Pushbutton
Switch

Figure 17-2. *The Amazing Pushbutton Fritzing diagram*

Tech Note

The clicking sound made when pressing the pushbutton switch provides audible feedback of closing electrical contacts.

Upload the Amazing Pushbutton Sketch

With the Amazing Pushbutton built, it's time to upload the sketch. Example 17-1 sends digital information to the Arduino IDE (integrated development environment) Serial Monitor and turns the onboard LED on and off with each press of the pushbutton switch. Here are the steps you'll need to follow:

1. Attach the Arduino to your computer using a USB cable.

2. Open the Arduino software and type Example 17-1 into the software's text editor.

3. Upload the sketch to the Arduino.

Example 17-1. *The Amazing Pushbutton sketch*

```
/*
 * The Amazing Pushbutton
 *
 * Reads a digital input from a pushbutton switch and sends the letter
 * L or H to the Serial Monitor.
 *
 *
 */


// variables for input pin and control LED
  int digitalInput = 7;
  int LEDpin = 13;

// variable to store the value
  int value = 0;

void setup(){

// declaration pin modes
  pinMode(digitalInput, INPUT);
  pinMode(LEDpin, OUTPUT);

// begin sending over serial port
  Serial.begin(9600);
}

void loop(){
// read the value on digital input
  value = digitalRead(digitalInput);

// write this value to the control LED pin
digitalWrite(LEDpin, value);

// if value is high then send the letter 'H'; otherwise, send 'L' for low
if (value)  Serial.print('H');
```

```
        else
        Serial.print('L');

   // wait a bit to not overload the port
     delay(10);
}
```

Once the Amazing Pushbutton sketch has been uploaded to the Arduino, the Serial Monitor will display "L" repeatedly in a row, as shown in Figure 17-3. Press the push-button switch, and the Serial Monitor displays "H" repeatedly in a row (see Figure 17-4).

Figure 17-3. *L's being displayed on the Arduino Serial Monitor*

Figure 17-4. *H's being displayed on the Arduino Serial Monitor*

Download and Install Processing Notes

Before building this awesome visual Arduino Microcontroller project, you have to install the Processing programming language on your computer. Here are the installation instructions:

1. Go to the Processing download web page (*https://processing.org/download/? processing*).
2. Select the software that meets your operating system's requirements.
3. Once the Processing software has been downloaded to your hard drive, follow the prompts to complete the installation process.

After installing the Processing programming language onto your computer, you're now ready to build the visualization software for the Amazing Pushbutton device!

Let's Visualize Digital Data with Processing

The characters "L" and "H" are an interesting way to represent the information you get when the pushbutton turns on and off. But if we really want to see the "magic" of the pushbutton, we'll need to use a graphical software language called

Processing. Processing software allows digital information (actually, just about any kind of information) to be changed into computer graphics quite easily.

With the Arduino attached to your computer, the state of the pushbutton, represented by L's and H's, can be changed to a colorful scale, showing increasing and decreasing numbers. The pa_Pushbutton Processing sketch displays an interactive scale on your computer screen, as shown in Figure 17-5.

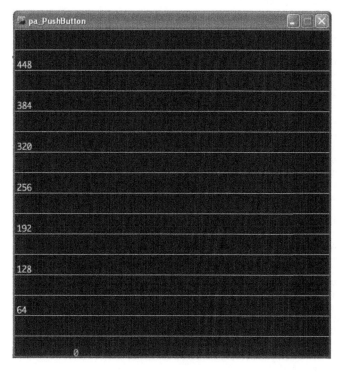

Figure 17-5. *We start with a very simple background scale created in Processing*

 Tech Note

Watch the online PBS Off Book episode titled "The Art of Data Visualization" (*https://www.youtube.com/watch?v=AdSZJzb-aX8*).

Example 17-2 shows the pa_Pushbutton Processing sketch for the Amazing Pushbutton.

Example 17-2. *The pa_Pushbutton Processing sketch*

```
/*
 *  pa_PushButton
```

```
*
*   Reads the values which represent the state of a pushbutton
*   from the serial port and draws increasing/decreasing horizontal lines.
*
*
*/

// importing the processing serial class
import processing.serial.*;

// the display item draws background and grid
  DisplayItems di;

// definition of window size and framerate
  int xWidth = 512;
  int yHeight = 512;
  int fr = 24;

// attributes of the display
  boolean bck = true;
  boolean grid = true;
  boolean g_vert = false;
  boolean g_horiz = true;
  boolean g_values = true;
  boolean output = false;

// variables for serial connection, portname, and baudrate have to be set
  Serial port;
  int baudrate = 9600;
  int value = 0;

// variables to draw graphics
  int actVal = 0;
  int num = 6;
  float valBuf[] = new float[num];
  int i;

// lets user control DisplayItems properties and value output in console
void keyPressed(){
  if (key == 'b' || key == 'B') bck=!bck; // background black/white
  if (key == 'g' || key == 'G') grid=!grid; // grid on/off
  if (key == 'v' || key == 'V') g_values=!g_values; // grid values on/off
  if (key == 'o' || key == 'O') output=!output; // turns value output on/off
}

void setup(){
  // set size and framerate
  size(xWidth, yHeight); frameRate(fr);

  // establish serial port connection
  // The "2" corresponds to the 3rd port (counting from 0) on the Serial
  // Port list dropdown. You might need to change the 2 to something else.
  String portname =Serial.list()[2];
  port = new Serial(this, portname, baudrate);
  println(port);
```

```
    // create DisplayItems object
    di = new DisplayItems();

    // clear value buffer
    for(i=0; i < num; i++) {
      valBuf[0] = 0;
    }

}

void drawPushButtonState(){
  // read through the value buffer
  // and shift the values to the left
  for(i=1; i < num; i++) {
    valBuf[i-1] = valBuf[i];
  }
  // add new values to the end of the array
  valBuf[num-1] = actVal;
  noStroke();
  // reads through the value buffer and draws lines
  for(int i=0; i < num; i=i+2) {
    fill(int((valBuf[i]*255)/height), int((valBuf[i]*255)/height) , 255);
    rect(0, height-valBuf[i], width, 3);
    fill(int((valBuf[i+1]*255)/height), 255, 0 );
    rect(0, height-valBuf[i+1], width, 3);
  }
  // display value
  fill(((bck) ? 185 : 75));
  text( ""+(actVal), 96, height-actVal);
}

void serialEvent(int serial){
  // if serial event is 'H' actVal is increased
  if(serial=='H') {
    actVal = (actVal < height - (height/16)) ?
            (actVal + int(actVal/(height/2))+1) :
            (actVal = height - (height/(height/2)));

    if (output)
      println("Value read from serial port is 'H' - actualValue is now "
              + actVal);
  } else {
    // if serial event is 'L' actVal is decreased
    actVal = (actVal > 1) ?
            (actVal = actVal - int(actVal/64)-1) :
            (actVal=0);
    if (output)
      println("Value read from serial port is 'L' - actualValue is now "
              + actVal);
  }
}

void draw(){
  // listen to serial port and trigger serial event
```

```
  while(port.available() > 0){
      value = port.read();
      serialEvent(value);
  }
  // draw background, then PushButtonState and
  // finally rest of DisplayItems
  di.drawBack();
  drawPushButtonState();
  di.drawItems();
}
```

Next, we need to use the DisplayItems sketch to display the interactions with the Arduino on your screen. To do this, you need to open a new tab in the Processing IDE for the DisplayItems sketch. Enter Example 17-3 into the new tab in the Processing IDE text editor.

Example 17-3. *The DisplayItems Processing sketch*

```
/*
 *  DisplayItems
 *
 *  This class draws background color, grid and value scale
 *  according to the boolean variables in the pa_file.
 *
 *  This file is part of the Arduino meets Processing Project.
 *  For more information visit http://www.arduino.cc.
 *
 *  created 2005 by Melvin Ochsmann for Malmo University
 *
 */

class DisplayItems{

// variables of DisplayItems object
PFont font;
int gridsize;
int fontsize = 10;
String fontname = "Monaco-14.vlw";
String empty="";
int i;

// constructor sets font and fontsize
DisplayItems(){
  font = loadFont(fontname);
  gridsize = (width/2)/16+(height/2)/16;
  if(gridsize > 20) fontsize = 14;
  if(gridsize > 48) fontsize = 22;
  textFont(font, fontsize);
}

// draws background
void drawBack(){
      background( (bck) ? (0) : (255)  );
}
```

```
// draws grid and value scale
void drawItems(){
  if(grid){   stroke( (bck) ? (200) : (64) );
              fill((bck) ? (232) : (32) );

  // vertical lines
  if(g_vert){
    for (i=0; i < width; i+=gridsize){
    line(i, 0, i, height);
    textAlign(LEFT);
    if (g_values &&
        i%(2*gridsize)==0 &&
        i < (width-(width/10)))
      text( empty+i, (i+fontsize/4), 0+fontsize);
  }}

  // horizontal lines
  if(g_horiz){
    for (int i=0; i < height; i+=gridsize){
    line(0, i, width, i);
    textAlign(LEFT);
    if (g_values &&
        i%(2*gridsize)==0)
      text( empty+(height-i), 0+(fontsize/4), i-(fontsize/4));
  }}
  }
 }
}// end class DisplayItems
```

After typing the sketch, click the play button to obtain the image shown in Figure 17-5. Press the Arduino's pushbutton to watch the numbers increase and the color bar (horizontal lines) move up the scale, as shown in Figure 17-6.

The Amazing Pushbutton Easter Eggs!

In addition to creating colorful lines and changing numbers, you can change the look of the scale using the following keys:

- b/B: toggles background black/white
- g/G: toggles grid on/off
- v/V: toggles grid values on/off
- o/O: turns value output on/off

See if you can find additional eggs hidden in the Processing sketch. Happy Hacking!

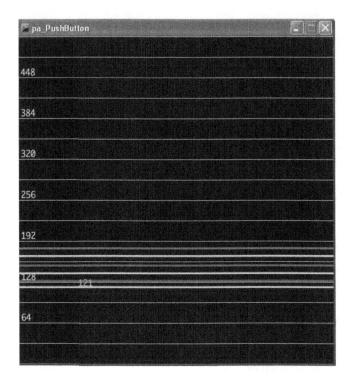

Figure 17-6. *The Amazing Pushbutton in action*

Tech Note

For additional information about processing, see Casey Reas and Ben Fry's *Getting Started with Processing* (Maker Media, 2010).

The block diagram in Figure 17-7 shows the electronic component blocks and the data flow for the Amazing Pushbutton. A Fritzing electronic circuit schematic diagram of the Amazing Pushbutton is shown in Figure 17-8.

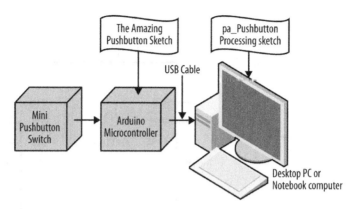

The Amazing Pushbutton Sketch

pa_Pushbutton Processing sketch

USB Cable

Mini Pushbutton Switch

Arduino Microcontroller

Desktop PC or Notebook computer

Figure 17-7. *The Amazing Pushbutton block diagram*

Troubleshooting Tips for Processing

As in all Maker projects, a bug can occasionally creep in. Processing is an awesome software package for developing cool Arduino microcontroller projects, but it can be challenging to use. Here are a few troubleshooting tips for the most common problems that can occur:

- Make sure the Arduino microcontroller is communicating with the Processing software through USB connection. If the Arduino is not attached to the Processing software, it may cause communication errors.

- Make sure the Amazing Pushbutton sketch is running on Arduino before starting the Processing sketch. If the Processing software is unable to obtain data from the Arduino microcontroller (because it wasn't running), it will generate an "unrecognized device error."

- Make sure text for both the Arduino and Processing sketches is typed correctly as shown in the software listings. Most of the software bugs are caused by syntax or incorrectly typed code for both programming languages.

Following these three guidelines should minimize your frustration when it comes to debugging the Amazing Pushbutton device project build.

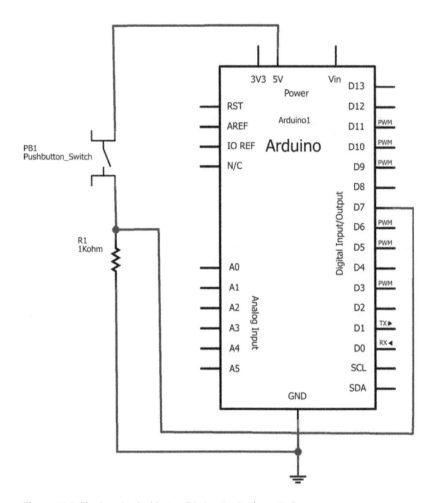

Figure 17-8. *The Amazing Pushbutton Fritzing circuit schematic diagram*

Something to Think About

How can the letters "L" and "H" in Figure 17-3 and Figure 17-4 be replaced with the numbers "0" and "1"?

The Terrific Tilt Switch (with Processing)

18

Processing is an awesome programming language that creates graphics and pictures that you can move in fun ways across the computer screen. Do you remember the tilt switch from Chapter 3? It was an electrical device capable of controlling electronic devices, based on its orientation or position. If you combine the Processing language with a tilt switch, you can create computer graphics that move across the screen when you make simple body gestures like waving a hand or raising and lowering an arm! In this project, a white circle will move from side to side on your computer screen as you rotate the tilt switch.

The electronic components to build this device are shown in the Parts List. The Terrific Tilt Switch is shown in Figure 18-1.

Parts List

- Arduino microcontroller
- MakerShield kit
- R1: 1KΩ resistor (brown, black, red stripes)
- S1: tilt switch
- USB cable

Figure 18-1. *The Terrific Tilt Switch*

Let's Build a Terrific Tilt Switch

The Terrific Tilt Switch, like the Amazing Pushbutton, requires a USB cable to send digital information from the switch to the computer screen. As shown in Figure 18-1, the device is quite simple to build: it requires just a 1KΩ fixed resistor and a tilt switch. The two components are connected in series like the Amazing Pushbutton device. Where the two components tie together, a jumper wire connects between them and pin D7 of the Arduino microcontroller.

The Terrific Tilt Switch can be built using the Fritzing wiring diagram shown in Figure 18-2. The placement of the parts is not critical, so have some fun placing the components in different places. Although the Fritzing diagram shows a mini breadboard, feel free to use the MakerShield protoboard if you want.

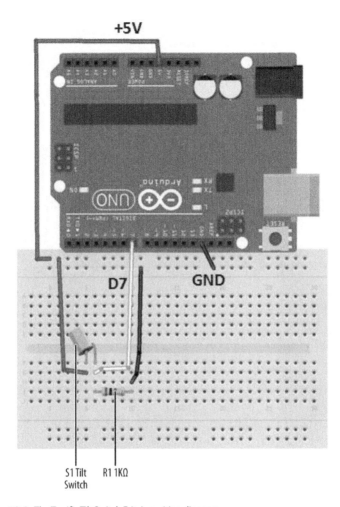

D7   GND

S1 Tilt   R1 1KΩ
Switch

Figure 18-2. *The Terrific Tilt Switch Fritzing wiring diagram*

Upload the Terrific Tilt Switch Sketch

It's time to upload the Arduino sketch for the Terrific Tilt Switch. Example 18-1 takes information from the tilt switch and sends it to the Arduino IDE (integrated development environment) Serial Monitor, displaying a series of the characters "H" and "L" with each rotation of the tilt switch.

Did you notice that parts of the program look like the listing shown in Chapter 17? That's because the serial communication technique—the part of the code that lets the Arduino talk with Processing—remains the same no matter what the Arduino is using as input or how Processing displays the data. Here are the steps you'll need to follow:

1. Attach the Arduino to your computer using a USB cable.

2. Open the Arduino software and type Example 18-1 into the software's text editor.

3. Upload the sketch to the Arduino.

Once the Terrific Tilt Switch sketch has been uploaded to the Arduino, the Serial Monitor will display "L" repeatedly in a row, as shown in Figure 18-3. If you tilt the switch, the Serial Monitor will display "H" repeatedly (see Figure 18-4).

Figure 18-3. *L's being displayed on the Arduino Serial Monitor*

Figure 18-4. *H's being displayed on the Arduino Serial Monitor*

Example 18-1. *The Terrific Tilt Switch sketch*

```
/*
* The Terrific Tilt Switch
*
* Reads a digital input from a tilt switch and sends a series of
* L's or H's to the Serial Monitor.
*
*
*/


// variables for input pin and control LED
  int digitalInput = 7;
  int LEDpin = 13;

// variable to store the value
 int value = 0;

void setup(){

// declaration pin modes
  pinMode(digitalInput, INPUT);
  pinMode(LEDpin, OUTPUT);

// begin sending over serial port
  Serial.begin(9600);
}
```

```
void loop(){
// read the value on digital input
  value = digitalRead(digitalInput);

// write this value to the control LED pin
digitalWrite(LEDpin, value);

// if value is high then send the letter 'H'; otherwise, send 'L' for low
if (value)  Serial.print('H');
            else
            Serial.print('L');

 // wait a bit to not overload the port
  delay(10);
}
```

Let's Visualize Digital Data with Processing

With the Arduino attached to a Processing sketch running on your computer, the digital information (L's and H's) from the Arduino can be changed to a horizontally moving white circle based on the orientation of the tilt switch, as shown in Figure 18-5 and Example 18-2.

Tech Note

Check out the Processing sketch listings for Arduino projects (*http://playground.arduino.cc/Interfacing/Processing*).

Example 18-2. *The pa_Tilt Processing sketch*

```
/*
 *  pa_Tilt
 *
 *  Reads the values which represent the state of a Tilt switch
 *  from the serial port and draws white-filled circle with vertical lines.
 *  created 2005 by Melvin Ochsmann for Malmo University
 *
 */

import processing.serial.*;

  DisplayItems di;

  int xWidth = 512;
  int yHeight = 512;
  int fr = 24;

  boolean bck = true;
  boolean grid = true;
  boolean g_vert = true;
```

```
boolean g_horiz = false;
boolean g_values = false;
boolean output = true;

Serial port;

// The "2" corresponds to the 3rd port (counting from 0) on the Serial
// Port list dropdown. You might need to change the 2 to something else.
String portname = Serial.list()[2];
int baudrate = 9600;
int value = 0;
boolean tilted = true;
float a = 0;
int speed = 5; // how many pixels that the circle will move per frame

void keyPressed(){

  if (key == 'b' || key == 'B') bck=!bck;
  if (key == 'g' || key == 'G') grid=!grid;
  if (key == 'v' || key == 'V') g_values=!g_values;
  if (key == 'o' || key == 'O') output=!output;
}

void setup(){

      size(xWidth, yHeight);
      frameRate(fr);

      di = new DisplayItems();

      port = new Serial(this, portname, baudrate);
      println(port);
}
// Method moves the circle from one side to another,
// keeping within the frame
void moveCircle(){

  if(tilted) {
      background(0);

      a = a + speed;
      if (a > (width-50)) {
        a = (width-50);
      }
      ellipse(a, (width/2), 100,100);

  }else{
      background(0);

      a = a - speed;
      if (a < 50) {
        a = 50;
      }
      ellipse(a, (width/2), 100,100);
```

```
        }
    }

void serialEvent(int serial){
        if(serial=='H') {
            tilted = true;
             if(output) println("High");

        }else {
            tilted = false;
                 if(output) println("Low");
         }
}

void draw(){

  while(port.available() > 0){
        value = port.read();
        serialEvent(value);
    }

    di.drawBack();

    moveCircle();

    di.drawItems();

}
```

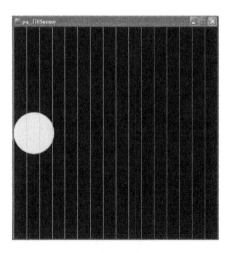

Figure 18-5. *An interactive (moving) white-filled circle created in Processing*

Tech Note

There are a couple of Easter eggs embedded in the pa_Tilt Processing sketch that will allow you to change the appearance of the display. Good Hunting!

Next, open a new tab in the Processing IDE and add Example 18-3. After typing the sketch, click the play button. Your computer screen should show something very similar to the image shown in Figure 18-5. Rotate the tilt switch and watch the white circle move across your computer screen, as shown in Figure 18-6.

Example 18-3. *The DisplayItems Processing sketch*

```
/*
 * DisplayItems
 *
 * This class draws background color, grid and value scale
 * according to the boolean variables in the pa_file.
 *
 * This file is part of the Arduino meets Processing Project.
 * For more information visit http://www.arduino.cc.
 *
 * created 2005 by Melvin Ochsmann for Malmo University
 *
 */

class DisplayItems{

// variables of DisplayItems object
PFont font;
int gridsize;
int fontsize = 10;
String fontname = "Monaco-14.vlw";
String empty="";
int i;

// constructor sets font and fontsize
DisplayItems(){
  font = loadFont(fontname);
  gridsize = (width/2)/16+(height/2)/16;
  if(gridsize > 20) fontsize = 14;
  if(gridsize > 48) fontsize = 22;
  textFont(font, fontsize);
}

// draws background
void drawBack(){
    background( (bck) ? (0) : (255) );
}

// draws grid and value scale
void drawItems(){
  if(grid){ stroke( (bck) ? (200) : (64) );
```

```
                fill((bck) ? (232) : (32) );

  // vertical lines
  if(g_vert){
    for (i=0; i < width; i+=gridsize){
    line(i, 0, i, height);
    textAlign(LEFT);
    if (g_values &&
        i%(2*gridsize)==0 &&
        i < (width-(width/10)))
      text( empty+i, (i+fontsize/4), 0+fontsize);
  }}

  // horizontal lines
  if(g_horiz){
    for (int i=0; i < height; i+=gridsize){
    line(0, i, width, i);
    textAlign(LEFT);
    if (g_values &&
        i%(2*gridsize)==0)
      text( empty+(height-i), 0+(fontsize/4), i-(fontsize/4));
    }}
  }
 }
}// end class Display
```

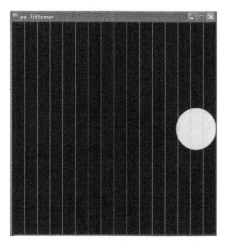

Figure 18-6. *The Terrific Tilt Switch in action: the white circle has moved to the right side of the screen*

 Tech Note

A class defines the look and operation of software objects.

The block diagram in Figure 18-7 shows the electronic component blocks and the data flow for the Terrific Tilt Switch. A Fritzing electronic circuit schematic diagram of the Terrific Tilt Switch is shown in Figure 18-8.

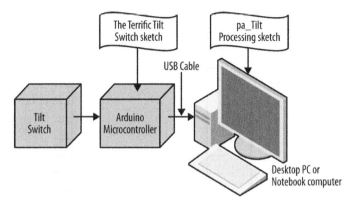

Figure 18-7. *The Terrific Tilt Switch block diagram*

Something to Think About

How can an external LED be wired to the MakerShield protoboard to visually represent the state of the tilt switch (just like the letters "L" and "H" do in the Serial Monitor)?

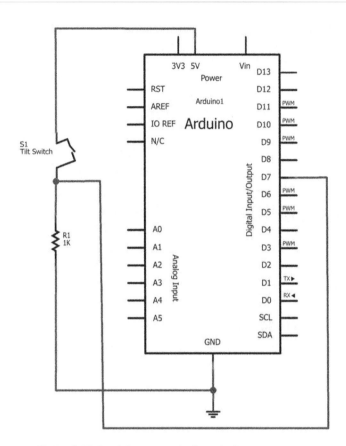

Figure 18-8. *The Terrific Tilt Switch Fritzing circuit schematic diagram*

The Rocket Launching Game (with Processing)

19

Would you like to build a rocket launching game using electronic components from the Ultimate Microcontroller Pack? How cool would it be to launch the rockets from your Maker bench or bedroom? Since launching *real* rockets is a bit beyond the scope of this book, we're going to use four pushbutton switches and the Arduino microcontroller to build a *virtual* rocket launcher. The rest is done in Processing. The electronic components to build this gadget are shown in the Parts List. The Rocket Launcher is shown in Figure 19-1.

Parts List

- Arduino microcontroller
- MakerShield kit
- R1-R5: 1KΩ resistors (brown, black, red stripes)
- S1-S4: pushbutton switches
- USB cable
- Full-size breadboard

Figure 19-1. *The Rocket Launcher*

Let's Build a Rocket Game

The Rocket Game, like the projects in Chapter 17 and Chapter 18, requires the use of a USB cable to send digital information from four pushbutton switches to the computer screen. As shown in Figure 19-1, the breadboard circuit is quite simple to build and requires five 1KΩ fixed resistors and four pushbutton switches.

The basic digital circuit consists of a pushbutton switch and resistor wired in series. This wiring connection is repeated three times for the remaining switches. These switches are connected to pins D3 through D7 of the Arduino microcontroller.

The Rocket Game can be built using the Fritzing wiring diagram shown in Figure 19-2. The placement of the parts is not critical, so experiment with the location of the various electronic components, and the overall wiring of the device. One challenge is to wire all of the electronic components using the awesome (but kind of small) MakerShield protoboard. Can you fit it all on there?

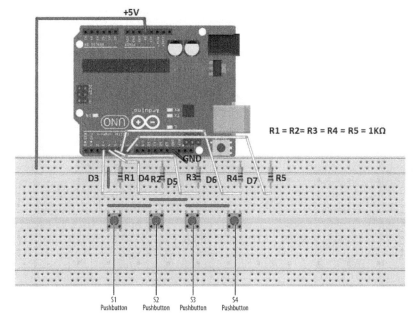

R1 = R2 = R3 = R4 = R5 = 1KΩ

Figure 19-2. *The Rocket Launcher Fritzing wiring diagram*

Tech Note

Processing language version 2.0 (*http://playground.arduino.cc/Inter facing/Processing*) is available for download.

Upload the MultiDigital4 Sketch

After building the Rocket Game pushbutton circuit and checking for wiring errors, it is time to upload the sketch. Example 19-1 sends digital information to the Arduino IDE (integrated development environment) Serial Monitor, displaying the numbers 0, 1, 2, and 4 with each individual press of the four pushbutton switches. The serial communication technique used in Chapter 17 and Chapter 18 remains the same for the Arduino software to talk with the Processing programming language. Here are the steps you'll need to follow:

1. Attach the Arduino to your computer using a USB cable.

2. Open the Arduino software and type Example 19-1 into the software's text editor.

3. Upload the sketch to the Arduino.

The Serial Monitor will start to display numbers, as shown in Figure 19-3, each time you press the pushbutton. Pushing various combinations of switches will show new results. This information will be displayed in Processing, and will create an awesome visual of rockets being launched on the screen.

Figure 19-3. *Decimal equivalent numbers being displayed on the Arduino Serial Monitor; pushbutton 3 has been pressed*

0, 1, 2, and ... 4?

Why are the pushbuttons producing the numbers 0, 1, 2, and 4? What happened to 3? The answer is that the buttons have been programmed to count in *binary*, a number system that is based on the powers of 2. Zero is just that, zero. Two raised to the 0th power equals 1. Two raised to the 1st power equals 2. Two raised to the 2nd power equals 4.

Why the preoccupation with powers of two? As computers have two primary states (+5 volts and 0 volts), it's easy to use those two states as the internal basis for everything the computer does. When computers need to count, add, or divide, they break the operation down into powers of two. In this project, instead of making the Arduino do the conversion, we've just started out using the powers of two.

Example 19-1. *The MultiDigital4 sketch*

```
/*
 *  MultiDigital4
 *
 *  Reads 8 digital inputs and sends their values over the serial port.
 *  A byte variable is used to store the state of all eight pins. This byte
 *  is then sent over the serial port.
 *
 *  modified ap_ReadDigital8 sketch by Melvin Oschmann
 *
 *  8 June 2013
 *  Don Wilcher
 *
 */

// 8 variables for each pin
   int digitalInput_1 = 3;
   int digitalInput_2 = 4;
   int digitalInput_3 = 5;
   int digitalInput_4 = 6;
   int digitalInput_5 = 7;
   int digitalInput_6 = 8;
   int digitalInput_7 = 9;
   int digitalInput_8 = 10;

// 8 variables to store the values
   int value_1 = 0;
   int value_2 = 0;
   int value_3 = 0;
   int value_4 = 0;
   int value_5 = 0;
   int value_6 = 0;
   int value_7 = 0;
   int value_8 = 0;

// byte variable to send state of all pins over serial port
   int myByte = 0;

// control LED
   int controlLED = 13;

void setup(){

// set pin modes
   pinMode(digitalInput_1, INPUT);  pinMode(digitalInput_2, INPUT);
   pinMode(digitalInput_3, INPUT);  pinMode(digitalInput_4, INPUT);
   pinMode(digitalInput_5, INPUT);  pinMode(digitalInput_6, INPUT);
   pinMode(digitalInput_7, INPUT);  pinMode(digitalInput_8, INPUT);

   pinMode(controlLED, OUTPUT);


// begin sending out over the serial port
```

```
  Serial.begin(9600);
}

void loop(){

// set 'myByte' to zero
  myByte = 0;

// then read all the INPUTS and store values
// in the corresponding variables
  value_1 = digitalRead(digitalInput_1);
  value_2 = digitalRead(digitalInput_2);

  value_3 = digitalRead(digitalInput_3);
  value_4 = digitalRead(digitalInput_4);

  value_5 = digitalRead(digitalInput_5);
  value_6 = digitalRead(digitalInput_6);

  value_7 = digitalRead(digitalInput_7);
  value_8 = digitalRead(digitalInput_8);

/* check if values are high or low and 'add' each value to myByte
 *   what it actually does is this:
 *
 *         00 00 00 00  ('myByte set to zero')
 *       | 00 10 10 00  ('3 and 5 are 1')
 *       --------------
 *         00 10 10 00  ('myByte after logical operation')
 *
 */

  if (value_1) {
    myByte = myByte | 0;
    digitalWrite(controlLED, HIGH);
  } else digitalWrite(controlLED, LOW);

  if (value_2) { myByte = myByte | 1; }
  if (value_3) { myByte = myByte | 2; }
  if (value_4) { myByte = myByte | 4; }
  if (value_5) { myByte = myByte | 8; }
  if (value_6) { myByte = myByte | 16; }
  if (value_7) { myByte = myByte | 32; }
  if (value_8) { myByte = myByte | 64; }

// send myByte out over serial port and wait a bit to not overload the port
  Serial.print(myByte);
  delay(10);
}
```

The Rocket Launcher with Processing

The numbers from the MultiDigital4 sketch will be interpreted by Processing and used to drive a cool graphics screen, with color numbers and text. The layout of the

Processing canvas is similar to the projects in Chapter 17 and Chapter 18 (with obvious differences in text and animation). After uploading the Rocket Game sketch to the Arduino, a jumbled blob of text and numbers along with a numbered grid will be displayed on your computer screen, as shown in Figure 19-4. If you look closely, you can see the word "rocket" repeated several times on the screen. Pressing pushbutton 1 will show the rocket launcher in action, as the text and associated number begin to rise on the numbered grid. Figure 19-5 shows an example of a virtual rocket being launched into the sky! Releasing the pushbutton allows the rocket to fall nicely back to earth.

Figure 19-4. *A blob of text and numbers*

Figure 19-5. *Rocket 3 being launched into the sky*

Another cool feature of the Rocket Game Processing sketch (Example 19-2) is the Console Monitor located below the numbered grid. The Console Monitor displays the binary status of the pushbuttons and launched rockets. As shown in Figure 19-5, one of the pushbuttons has a binary status of 1, while the other three pushbuttons show a binary status of 0. From that, can you deduce which pushbutton has been pressed?

The Console Monitor can also be used as a sketch debugging tool when developing graphics, animation, and Arduino applications.

Tech Note

The Processing programming language allows text-based information to be displayed using a Console Monitor. The canvas is used to display graphics and animation information.

Example 19-2. *The Rocket Game Processing sketch*

```
/*
 * The Rocket Game
 *
 *  Reads the values which represent the state of 4 switches
 *  from the serial port and draws a graphical representation.
 *  Sketch inspired by Melvin Ochsmann's Multiple8 Switches
 *
 *  05 June 2013
 *  modified by Don Wilcher
 */

// importing the processing serial class
import processing.serial.*;

// the display item draws the background and grid
  DisplayItems di;

// definition of window size and framerate
  int xWidth = 512;
  int yHeight = 512;
  int fr = 12;

// attributes of the display
  boolean bck = true;
  boolean grid = true;
  boolean g_vert = false;
  boolean g_horiz = true;
  boolean g_values = true;
  boolean output = true;

// variables for serial connection, port name, and baud rate have to be set
  Serial port;

  // establish serial port connection
  // The "2" corresponds to the 3rd port (counting from 0) on the Serial
  // Port list dropdown. You might need to change the 2 to something else.
  String portname =Serial.list()[2];
  int baudrate = 9600;
  int value = 0;

 // variables to draw graphics
  int i;

  // if you would like to change fonts, make sure the font file (which
  // can be created with Processing) is in the data directory
  String fontname2 = "Helvetica-Bold-96.vlw";
```

```
    int fontsize2 = 72;   // change size of text on screen
    PFont font2;
    float valBuf[] = new float[8];
    int xpos, ypos;

// lets user control DisplayItems properties and value output in console
void keyPressed(){
    if (key == 'b' || key == 'B') bck=!bck;  // background black/white
    if (key == 'g' || key == 'G') grid=!grid;  // grid on/off
    if (key == 'v' || key == 'V') g_values=!g_values;  // grid values on/off
    if (key == 'o' || key == 'O') output=!output; //turns value output on/off
}

void setup(){
    // set size and framerate
    size(xWidth, yHeight); frameRate(fr);
    // establish serial port connection
    port = new Serial(this, portname, baudrate);
    println(port);
    // create DisplayItems object
    di = new DisplayItems();
    // load second font for graphical representation and clear value buffer
    font2 = loadFont(fontname2);
    for(i = 0; i < valBuf.length; i++ ){
        valBuf[i] = (height/2);
    }
}

void drawFourSwitchesState(){
    textFont(font2, fontsize2);
    if (output) print("4Switches Statuses: ");

    // takes value, interprets it as a byte
    // and reads each bit
    for (i=0; i < 4 ; i++){

        if(output) print(value & 1);
        print("ROCKET!");

        // if a bit is 1, increase the corresponding value in value buffer
        // array by 1
        if ( (value & 1) == 1){  // if 0, number drops when pushbutton is
                                 // pressed; if 1, number goes up when
                                 // pushbutton is pressed

            if(valBuf[i] > fontsize2 ) valBuf[i] -=1;
            // if a bit is 0, decrease corresponding value
        }else{
            if(valBuf[i] < height) valBuf[i] += 1;
        }

        if(output)
            print(".");

        // draw number for each value at its current height
```

```
      fill( ( (i%3==0) ? 255 : 0 ),
            ( (i%3==1) ? 255 : 0 ) ,
            ( (i%3==2) ? 255 : 0 ));
      text( ( "ROCKET"+(i+1) ),
            (i*(width/12)) +  (width/15),
              valBuf[l]); // prints "ROCKET" along with number
      value = value >> 1;

  } // end for loop
  if(output)
    println("");
}

void draw(){
  // listen to serial port and set value
  while(port.available() > 0){
      value = port.read();
  }
  // draw background, then four switches and finally rest of DisplayItems
  di.drawBack();
  drawFourSwitchesState();
  di.drawItems();
}
```

Next, the DisplayItems sketch is required for the interaction of the Rocket Game and the computer graphics to be visible on your computer screen. Enter the DisplayItems sketch shown in Example 19-3 into the Processing IDE text editor. Note that a second tab needs to be inserted within the IDE for the DisplayItems sketch. After typing the sketch, click the play button to obtain the image shown in Figure 19-4. Press a push-button on the Arduino breadboard and watch the rocket move up your computer screen, as shown in Figure 19-5.

Tech Note

There are a couple of Easter eggs embedded in the Rocket Game Processing sketch that allow you to change the appearance of the display. Also, the onboard LED turns on with one of the pushbuttons. Good Hunting!

Example 19-3. *The DisplayItems Processing sketch*

```
/*
 *  DisplayItems
 *
 *  This class draws background color, grid and value scale
 *  according to the boolean variables in the Rocket Launcher file.
 *
 *  This file is part of the Arduino meets Processing Project.
 *  For more information visit http://www.arduino.cc.
 *
 *  created 2005 by Melvin Ochsmann for Malmo University
 *
```

```
*/

class DisplayItems{

// variables of DisplayItems object
PFont font;
int gridsize;
int fontsize = 10;
String fontname = "Monaco-14.vlw";
String empty="";
int i;

// constructor sets font and fontsize
DisplayItems(){
  font = loadFont(fontname);
  gridsize = (width/2)/16+(height/2)/16;
  if(gridsize > 20) fontsize = 14;
  if(gridsize > 48) fontsize = 22;
}

// draws background
void drawBack(){
      background( (bck) ? (0) : (255)  );
}

// draws grid and value scale
void drawItems(){
  textFont(font, fontsize);

  if(grid){  stroke( (bck) ? (200) : (64) );
             fill((bck) ? (232) : (32) );

  // vertical lines
  if(g_vert){
    for (i=0; i < width; i+=gridsize){
    line(i, 0, i, height);
    textAlign(LEFT);
    if (g_values &&
        i%(2*gridsize)==0
        && i < (width-(width/10)))
      text( empty+i, (i+fontsize/4), 0+fontsize);
  }}

  // horizontal lines
  if(g_horiz){
    for (int i=0; i < height; i+=gridsize){
    line(0, i, width, i);
    textAlign(LEFT);
    if (g_values &&
        i%(2*gridsize)==0)
      text( empty+(height-i), 0+(fontsize/4), i-(fontsize/4));
    }}
  }
 }
}// end class Display
```

Tech Note

The size of the letters can be changed with the `fontsize` variable.

The block diagram in Figure 19-6 shows the electronic component blocks and the data flow for the Rocket Game. A Fritzing electronic circuit schematic diagram of the gadget is shown in Figure 19-7. Electronic circuit schematic diagrams are used by electrical/electronic engineers to design and build cool interactive electronic products for society.

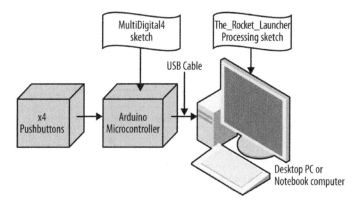

Figure 19-6. *The Rocket Game block diagram*

Something to Think About

How can the word "ROCKET" be replaced with "FIRE" within the Rocket Game Processing sketch?

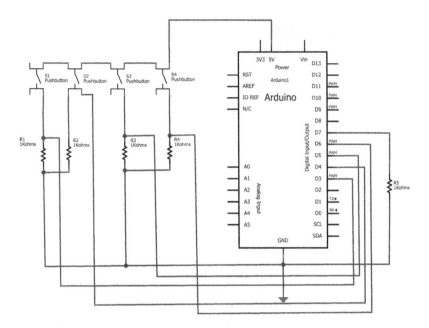

Figure 19-7. *The Rocket Game Fritzing circuit schematic diagram*

Temperature Indicator (with Processing)

Here's an awesome project that allows you to quickly check the temperature of the environment using a few electronic components from the Ultimate Microcontroller Pack. This temperature indicator uses a small electronic sensor called a thermistor, which changes its resistance—the way electricity flows through it—depending on the temperature. The electronic components to build this gadget are shown in the Parts List. The Temperature Indicator is shown in Figure 20-1.

Parts List

- Arduino microcontroller
- MakerShield kit
- R1: thermistor (green or black candy drop electronic component)
- R2: 10KΩ resistor (brown, black, orange stripes)
- USB cable
- LED1: red LED

Figure 20-1. *The Temperature Indicator*

Let's Build a Temperature Indicator

As shown in Figure 20-1, the breadboard analog circuit is quite simple to build, and requires only a thermistor and a 10KΩ fixed resistor wired in series. Where the two components are tied together, a jumper wire connects between them and pin A3 of the Arduino microcontroller.

The Temperature Indicator can be built using the Fritzing wiring diagram shown in Figure 20-2. Since there are only two electronic components, you have plenty of room for electrical wiring and breadboard placement of the components. Although the Fritzing wiring diagram shows a small breadboard, you can alternatively use the MakerShield protoboard to build the Temperature Indicator.

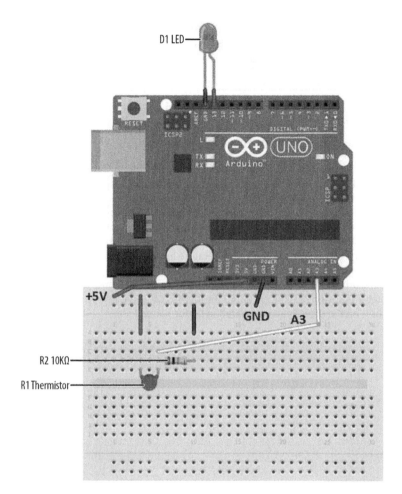

Figure 20-2. *The Temperature Indicator Fritzing wiring diagram*

Tech Note

A thermistor is a special variable resistor that changes its resistance based on temperature.

Upload the Temperature Indicator Sketch

After building the Temperature Indicator circuit and checking for wiring errors, it is time to upload the sketch. Example 20-1 sends analog information to the Arduino IDE (integrated development environment) Serial Monitor, and displays a series of numbers based on the thermistor's change in resistance. It uses the same serial communication technique used in Chapters 17, 18, and 19 to talk with the Processing programming language. Here are the steps you'll need to follow:

1. Attach the Arduino microcontroller to your computer using a USB cable.

2. Open the Arduino software and type Example 20-1 into the software's text editor.

3. Upload the sketch to the Arduino microcontroller.

With the Temperature Indicator sketch uploaded to the Arduino microcontroller, the Serial Monitor will display decimal numbers as shown in Figure 20-3. If you touch the thermistor—making it hotter with your own body heat—the Serial Monitor numbers will change. Also, if you add an external LED between pins D13 and GND, you'll have a visual indicator of when the thermistor's temperature has exceeded the threshold value programmed in the sketch. Figure 20-4 shows the Temperature Indicator's LED in operation. The Temperature Indicator is not an actual electronic thermometer but a device that can sense a certain heat level and respond to it by turning on an LED. The temperature units of Fahrenheit or Celsius are not displayed, thereby removing the concern about the thermistor's temperature resolution so the focus is on the device's actual operating performance.

Figure 20-3. *Decimal numbers being displayed on the Arduino Serial Monitor*

Figure 20-4. *Temperature Indicator detecting heat from a notebook computer.*

 Tech Note
The thermistor is used in electronic thermometers to measure temperature.

Example 20-1. *The Temperature Indicator sketch*

```
/*
 * Temperature_Indicator
 *
 * Reads an analog input from the input pin and sends the value
 * followed by a line break over the serial port. Data can be viewed
 * using the Serial Monitor.
 *
 * This file is part of the Arduino meets Processing Project:
 * For more information visit http://www.arduino.cc.
 *
 * created 2005 ap_ReadAnalog by Melvin Ochsmann for Malmo University
 *
 * 10 June 2013
 * modified by Don Wilcher
 *
 */

// variables for input pin and control LED
```

```
int analogInput = 3;
int LEDpin = 13;

// variable to store the value
  int value = 0;

// a threshold to decide when the LED turns on
  int threshold = 800;

void setup(){

// declaration of pin modes
  pinMode(analogInput, INPUT);
  pinMode(LEDpin, OUTPUT);

// begin sending over serial port
  Serial.begin(9600);
}

void loop(){
// read the value on analog input
  value = analogRead(analogInput);

// if value greater than threshold turn on LED
if (value < threshold) digitalWrite(LEDpin, HIGH);
else digitalWrite(LEDpin, LOW);

// print out value over the serial port
  Serial.println(value);

// and a signal that serves as separator between two values
  Serial.write(10);

// wait for a bit to not overload the port
  delay(100);
}
```

The Negative Temperature Coefficient (NTC) Sensor with Processing

When we connect the Temperature Indicator sketch to Processing, the thermistor temperature data from the sketch will be displayed in the Processing IDE Console Monitor, as well as on the main screen of the computer. The layout of this Processing canvas is simple. The graphics consist of two rectangular boxes with fluttering horizontal lines. The fluttering lines represent the thermistor's temperature, received from the Arduino microcontroller. An example of the fluttering lines and Console Monitor thermistor data is shown in Figure 20-5 and Figure 20-6. The NTC Sensor sketch is shown in Example 20-2. After uploading the NTC Sensor sketch to the Arduino microcontroller, two rectangular boxes with fluttering horizontal lines representing thermistor data will be visible on the computer screen.

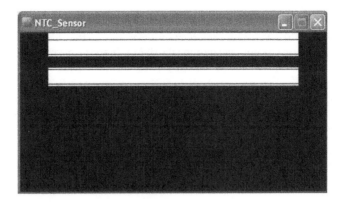

Figure 20-5. *Fluttering horizontal data lines*

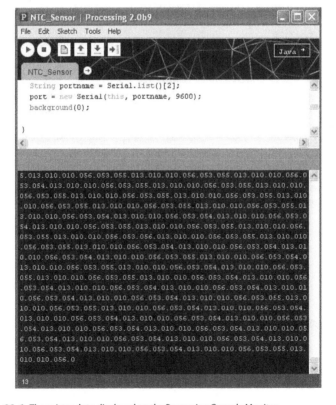

Figure 20-6. *Thermistor data displayed on the Processing Console Monitor*

Tech Note

The thermistor information being transmitted from the Arduino microcontroller through the USB cable and received by the Processing Console Monitor is a good example of data communications.

Example 20-2. *The NTC Sensor Processing sketch*

```
import processing.serial.*;

Serial port; // the Serial Port object is created
float val;    // variable used to receive thermistor data from Arduino



void setup() {
  size(440, 220); // size of canvas
  frameRate(30); // how fast the horizontal lines will flutter
  smooth(); // reduce jittering of the fluttering horizontal lines

  // The "2" corresponds to the 3rd port (counting from 0) on the Serial
  // Port list dropdown. You might need to change the 2 to something else.
  String portname = Serial.list()[2];
  port = new Serial(this, portname, 9600); // baud rate for COM port
  background(0); // create a black canvas

}

void draw() {
  if (port.available() > 0){ // check for available data on COM port
    val= port.read(); // store COM port data in variable "val"
    print(val); // print COM data on Console Monitor
    // val = map(val, 0, 255, 0, height);
    // float targetVal = val;
    // easedVal += (targetVal - easedVal)* easing;

  }

  rect(40, val, 360, 20); // display data has a fluttering horizontal
                          // line inside a rectangle

}
```

Tech Note

You can play with the horizontal line display rate by modifying the `frameRate(30)` processing instruction.

The block diagram in Figure 20-7 shows the electronic component blocks and the data flow for the Temperature Indicator. A Fritzing electronic circuit schematic diagram of the Temperature Indicator is shown in Figure 20-8. Electrical/electronic engineers use circuit schematic diagrams to design, build, and test cool interactive electronic products for society.

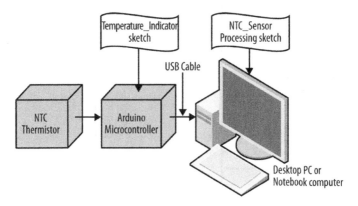

Figure 20-7. *The Temperature Indicator block diagram*

Something to Think About

How can a second LED be wired to the Arduino microcontroller to display when the temperature falls *below* a certain threshold?

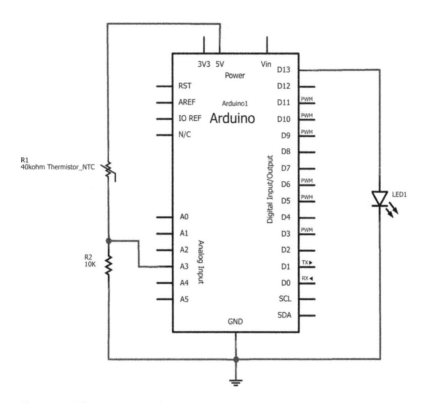

Figure 20-8. *The Temperature Indicator Fritzing circuit schematic diagram*

Sweeping Servo

21

An Electrical Motor Tester

Chapter 3 has an awesome project that changes a servo motor's angle of rotation based on the orientation of a tilt sensor. In that project, rotating the tilt sensor to either 0° or 90° positions will make the servo motor's shaft turn CW (clockwise) or CCW (counterclockwise). Also, the tilt sensor, along with the servo motor, can be used to build a simple animatronic controller for robotic puppets.

In this chapter, you can build a cool electrical tester to quickly test the limits of the Ultimate Microcontroller Pack servo motors, or those you may obtain outside the kit, like from a local Makerspace's electrical/mechanical parts bin.

The electronic components to build this electrical tester are shown in the Parts List. The Sweeping Servo Motor Tester is shown in Figure 21-1.

Parts List

- Arduino microcontroller
- MakerShield kit
- M1: DC servo motor

Figure 21-1. *The Sweeping Servo Motor Tester*

Let's Build a Servo Motor Tester

The Servo Motor Tester is quite simple to build and only requires the three components shown in the Parts List. With this tester, you will be able to quickly check any of the small voltage-based servo motors you may have in your junk box. The Servo Motor Tester can be built using the Fritzing wiring diagram shown in Figure 21-2. Since the major component for this project is the servo motor, placement of the parts on the breadboard is not critical. You have lots of room to explore different ways to place the servo motor when laying out the circuit on the breadboard.

In addition, by inserting the appropriate size solid wires into the three-pin female connector, you can easily make a male connecting component. This homebrew male connector makes it easy to insert the servo motor into a breadboard. (For further reference on building a servo motor male connector, see Figure 3-4 in Chapter 3.) Although the Fritzing wiring diagram shows a small breadboard, you can also use the MakerShield protoboard to build the Servo Motor Tester.

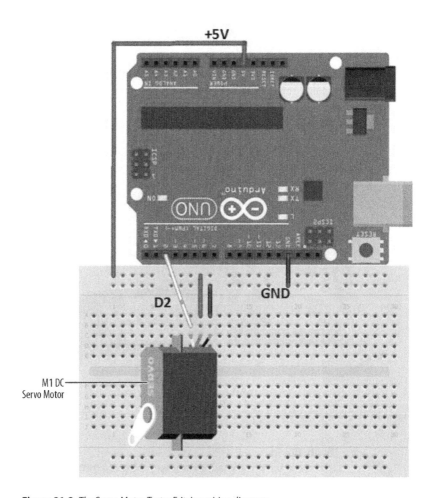

Figure 21-2. *The Servo Motor Tester Fritzing wiring diagram*

Tech Note
The color-coded wires for the Ultimate Microcontroller Pack are yellow (to D2), red (to +5V), and brown (to GND).

Upload the Sweeping Sketch

With the Servo Motor Tester built on the breadboard, now it's time to upload an Arduino sketch. Before uploading the sketch, check for wiring errors and make sure the servo motor connector is correctly attached to the breadboard. Example 21-1 sends a series of electrical pulses from the Arduino microcontroller's digital pin D9 to the servo motor. Here are the steps you'll need to follow:

1. Attach the Arduino microcontroller to your computer using a USB cable.

2. Open the Arduino software and type Example 21-1 into the software's text editor.

3. Upload the sketch to the Arduino microcontroller.

With the Sweeping sketch uploaded to the Arduino microcontroller, the servo motor will begin rotating CW and CCW continuously. Figure 21-3 shows a servo motor being tested by the Sweeping sketch.

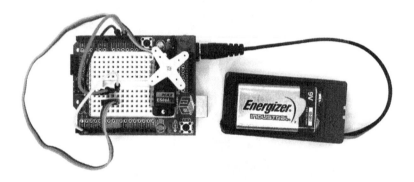

Figure 21-3. *A servo motor being tested using the Sweeping sketch*

 Tech Note

The size of the pulse width determines the servo motor's angle of rotation.

Example 21-1. *The Sweeping sketch*

```
#include <Servo.h>

Servo myservo;  // create servo object to control a servo
                // a maximum of eight servo objects can be created

int pos = 0;    // variable to store the servo position

void setup()
{

  myservo.attach(2);  // attaches the servo on pin 2 to the servo object
}
```

```
void loop()
{
  for(pos = 0; pos < 170; pos += 1)   // goes from 0 degrees to 170 degrees
  {                                    // in steps of 1 degree
    myservo.write(pos);                // move to position in variable 'pos'
    delay(15);                         // waits 15ms to reach the position
  }
  for(pos = 170; pos>=1; pos-=1)       // goes from 170 degrees to 0 degrees
  {
    myservo.write(pos);                // move to position in variable 'pos'
    delay(15);                         // waits 15ms to reach the position
  }
}
```

One final point to make about the Sweeping sketch: with some servo motors, a 180° pulse may cause the gears to grind. Therefore, experiment (gently!) with this value to learn the maximum CW and CCW rotation of your particular servo, without grinding the servo motor's gears.

 Tech Note

In Chapter 5 of *Make: Electronics*, there is a nice reference page describing various DC motors, including the servo motor.

The block diagram in Figure 21-4 shows the electronic component blocks and the data flow. The equivalent Fritzing electronic circuit schematic diagram of the Servo Motor Tester is shown in Figure 21-5.

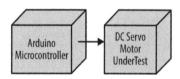

Figure 21-4. *The Servo Motor Tester block diagram*

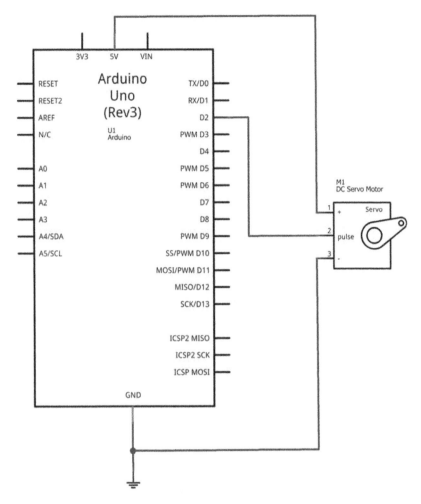

Figure 21-5. *The Servo Motor Tester Fritzing circuit schematic diagram*

Something to Think About

How can an LED be wired to the Arduino microcontroller to light up when the servo motor is at 180°?

Electronic Cricket

22

Temperature Sensing

Did you know that you can tell the temperature from crickets? Crickets chirp faster when the temperature is warmer, and they chirp more slowly when the air is cooler. Let's use our Arduino microcontroller skills and our knowledge about thermistors (Chapter 21) to build an electronic cricket. It won't *look* much like an insect, but the Arduino cricket will respond to temperature much the same way a real cricket does —its chirp will speed up as the air temperature warms up, and slow down as the air gets cooler. And unlike real crickets (which stop chirping when the temperature gets below about 60°F), your Arduino will continue to chirp almost to freezing!

The components to build this electronic cricket are shown in the Parts List. The Electronic Cricket is shown in Figure 22-1.

Parts List

- Arduino microcontroller
- MakerShield kit
- R1: negative temperature coefficient (NTC) thermistor (green or black candy drop; part number 503)
- R2: 10KΩ resistor (brown, black, orange stripes)
- R3: 1KΩ resistor (brown, black, red stripes)
- R4: 10K potentiometer
- SPKR1: 8Ω mini speaker

Figure 22-1. *The assembled Electronic Cricket*

Let's Build an Electronic Cricket

The Electronic Cricket is a creative, interactive device that produces electronic sounds using an Arduino microcontroller, a temperature sensor, two fixed resistors, a potentiometer, and a mini speaker. The values for these electronic components are in the Parts List. Follow the Fritzing wiring diagram shown in Figure 22-2 to construct the cricket.

When the project is built, you can immediately test the cricket by holding the temperature sensor between your fingers. The pitch of the sound coming out of the speaker, as well as the frequency of the chirping, will increase as the temperature rises. You can control the volume of the chirping with the potentiometer.

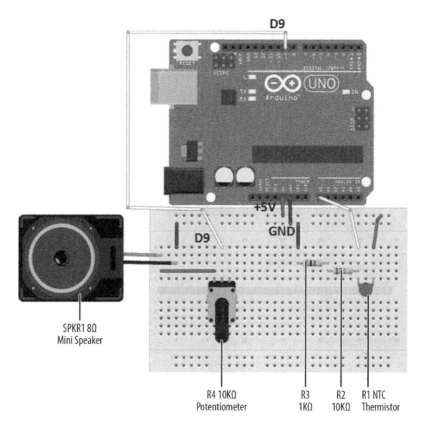

D9

+5V

D9

GND

SPKR1 8Ω
Mini Speaker

R4 10KΩ
Potentiometer

R3
1KΩ

R2
10KΩ

R1 NTC
Thermistor

Figure 22-2. *Electronic Cricket Fritzing wiring diagram.*

Tech Note

To convert cricket chirps to degrees Fahrenheit, you simply count the number of chirps in 14 seconds, then add a constant of 40 to get the temperature.

Example: 40 chirps per 14 seconds + 40 = 80°F

Upload the Electronic Cricket Sketch

With the Electronic Cricket wired on the breadboard, now it's time to upload the Arduino sketch. Before uploading the sketch, check for wiring errors on the breadboard. Example 22-1 sends a series of electrical pulses from the digital pin D9 on the Arduino microcontroller to the 8Ω mini speaker. Here are the steps you'll need to follow:

1. Attach the Arduino microcontroller to your computer using a USB cable.

2. Open the Arduino software and type Example 22-1 into the software's text editor.

3. Upload the sketch to the Arduino microcontroller.

The Arduino microcontroller is now programmed with the Electronic Cricket sketch. If everything is wired correctly, the 8Ω mini speaker should start to chirp somewhat like a real cricket. Touch the temperature sensor with your finger, and the chirping should speed up and get higher in pitch. Remove your finger from the temperature sensor and the chirping will decrease in speed and pitch.

The speed of the chirping per minute is approximately equal to the degrees in Fahrenheit. If you hear 80 chirps in one minute, the temperature is approximately 80° F. (If you don't want to wait a whole minute, simply count the chirps for 15 seconds and multiply by 4.)

Tech Note

For additional information about the tone( ) instruction, go to the reference website for the Arduino (*http://arduino.cc/en/Reference/HomePage*).

Example 22-1. *Electronic Cricket sketch*

```
/*
  Electronic Cricket

  Plays a pitch that changes based on a changing analog (temperature) input

*/

int expectedMax = 859;
int expectedMin = 330;
float fahrenheit = 0.0;

void setup() {
  // initialize serial communications (for debugging only):
  Serial.begin(9600);
}

void loop() {
  // read the sensor:
  int sensorReading = analogRead(A0);

  // print the sensor reading so you know its range
  Serial.println(sensorReading);

  // map the sensor analog input range
  // to the output pitch range (10 - 100Hz)
  // change the minimum and maximum input numbers below
```

```
// depending on the range your sensor's giving:


int thisPitch = map(sensorReading, expectedMin, expectedMax, 10, 100);
int thisTemperature =
  map(sensorReading, expectedMin, expectedMax, -10, 40);

fahrenheit = (5.0/9.0) * (thisTemperature + 32);

Serial.println(fahrenheit);

// play the pitch twice, to imitate a cricket:
tone(9, thisPitch, 10);

// the delay is proportional to the temperature
// (faster chirps mean higher temperatures)
delay(60000/(fahrenheit * 1000));

tone(9, thisPitch, 10);

delay(1);        // delay in between reads for stability
}
```

As shown in the sketch, the Electronic Cricket's pitch range can create fun electronic sounds easily by changing the first two numbers in the map( ) instruction. Play with the pitch to see how many cool electronic sounds you can create from your Electronic Cricket!

Tech Note
To understand how the potentiometer operates the mini speaker, here's a cool experiment (*http://www.allaboutcircuits.com/vol_6/ chpt_3/7.html*) you can try with the Ultimate Microcontroller Pack.

The block diagram in Figure 22-3 shows the electronic component blocks and the data flow for the Electronic Cricket. Also, the equivalent Fritzing electronic circuit schematic diagram of the Electonic Cricket is shown in Figure 22-4. Circuit schematic diagrams are used by electrical/electronic engineers and technicians to design, build, and test cool interactive electronic products for games, testing equipment, robots, and automobiles.

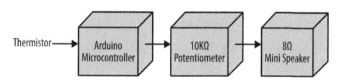

Figure 22-3. *The Electronic Cricket block diagram*

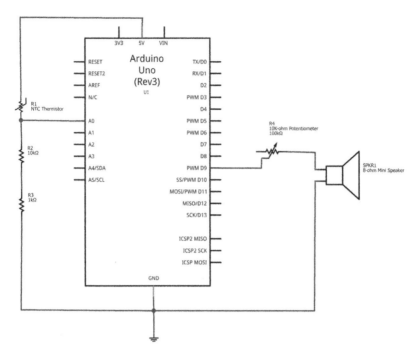

Figure 22-4. *The Electronic Cricket Fritzing circuit schematic diagram*

Something to Think About

How can the mini speaker be replaced with an LED for a visual pitch indicator?

A Pocket Stage Light

Temperature Sensing (Part 2)

Stage lighting provides a cool effect for concerts and plays. The various color lenses placed on the lights are used to complement the mood of the performers on stage. How awesome would it be to have an electronic gadget that can project colors on a wall without using special lenses and light bulbs? You'll be able to perform a magnificent lighting show for your family, friends, and local Makerspace members anywhere at any time. Also, the pocket lighting device can be operated by touch instead of pushbutton switches. With electronic parts obtained from the Ultimate Microcontroller Pack, you can build your own Pocket Stage Light. The electronic components to build the Pocket Stage Light are shown in the Parts List. The assembled Pocket Stage Light is shown in Figure 23-1.

Parts List

- Arduino microcontroller
- MakerShield kit
- R1: negative temperature coefficient (NTC) thermistor (green or black candy drop; part number 503)
- R2: 10KΩ resistor (brown, black, orange stripes)
- R3: 1KΩ resistor (brown, black, red stripes)
- R4: 330Ω resistor (orange, orange, brown stripes)
- LED1: RGB LED

Figure 23-1. *The assembled Pocket Stage Light*

Let's Build a Pocket Stage Light

Operating an electronic gadget with sensors is called physical computing. Other examples of physical computing devices are Microsoft's Kinect and smartphone touch screens. The Pocket Stage Light is operated by warm temperature. The temperature value is changed to an electrical voltage and used by the Arduino microcontroller to turn on an RGB LED. Control over the color lighting sequence of red, green, and blue is provided by an Arduino sketch.

Use the Fritzing wiring diagram shown in Figure 23-2 to build the Pocket Stage Light. Touching the thermistor with your finger will signal the Arduino microcontroller to turn on the red, green, and blue colors of the LED in sequence. After you release the thermistor, the color LEDs will continue to sequence for approximately 10 seconds.

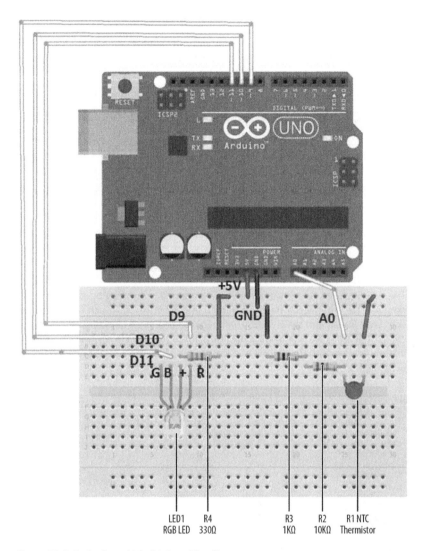

Figure 23-2. *Pocket Stage Light Fritzing wiring diagram*

Before uploading the sketch in Example 23-1 to the Arduino, check and correct any wiring errors on your breadboard using the Fritzing diagram shown in Figure 23-2.

Tech Note
With the right color mix, you can create a white light with an RGB LED. Check out Wikipedia's page on the RGB color model (*http://en.wikipedia.org/wiki/RGB_color_model*).

Upload the Pocket Stage Light Sketch

With the Pocket Stage Light wired on the breadboard, now it's time to upload the Arduino sketch. Example 23-1 turns on three Arduino microcontroller digital pins (D9, D10, and D11) in sequence that operate the red, green, and blue portion of the RGB LED. Here are the steps you'll need to follow:

1. Attach the Arduino microcontroller to your computer using a USB cable.

2. Open the Arduino software and type Example 23-1 into the software's text editor.

3. Upload the sketch to the Arduino microcontroller.

The Arduino microcontroller is now programmed with the Pocket Stage Light sketch. When the sketch starts running, the RGB LED is off. Touch the thermistor with your finger, and the RGB LED will begin to sequence its red, green, and blue colors. Releasing the sensor will allow the color sequencing to continue for approximately one second. Figure 23-3 and Figure 23-4 show the operation of the Pocket Stage Light.

Figure 23-3. *Pocket Stage Light projecting a green light on a whiteboard*

Figure 23-4. *Pocket Stage Light projecting a blue light on a whiteboard*

Example 23-1. *Pocket Stage Light sketch*

```
/*
  Pocket Stage Light
  The RGB LED will sequence in colors (blue, green, red) by use
  of a thermistor.


  15 August 2013
  by Don Wilcher

 */

int tsensorPin = A0;      // select the input pin for the temperature sensor
int RPin = 11;            // select the pin for the red LED
int GPin = 10;            // select the pin for the green LED
int BPin = 9;             // select the pin for the blue LED
int tsensorValue = 0;     // to store the value from the temperature sensor

void setup() {
  // declare the LED pins as outputs:
  pinMode(RPin, OUTPUT);
  pinMode(GPin, OUTPUT);
  pinMode(BPin, OUTPUT);
  Serial.begin(9600);
```

```
}

void loop() {
  // read the value from the sensor:
  tsensorValue = analogRead(tsensorPin);
  Serial.println(tsensorValue);
  delay(100);
  if (tsensorValue > 190){
  // turn the blue LED on:
  digitalWrite(BPin, LOW);
  digitalWrite(RPin, HIGH);
  // delay blue LED for 5 seconds:
  delay(5000);
  // turn the green LED on:
  digitalWrite(BPin, HIGH);
  digitalWrite(GPin, LOW);
  // delay green LED for 5 seconds:
  delay(5000);
  // turn the red LED on:
  digitalWrite(GPin, HIGH);
  digitalWrite(RPin, LOW);
  //delay red LED for 5 seconds:
  delay(5000);
  }
  else{
    // turn blue, green, and red LEDs off:
    digitalWrite(BPin, HIGH);
    digitalWrite(GPin, HIGH);
    digitalWrite(RPin, HIGH);
  }
}
```

You can see the thermistor's output on the Arduino's Serial Monitor. Also, as discussed in Chapter 22, you can experiment with a 10KΩ thermistor to see a difference in the RGB LEDs response. Observe the operation of the RGB LED closely to see a different turn-on response based on the 10KΩ thermistor component. Last, change the color sequence of the LEDs from the order listed within the sketch! Remember to record your software and electrical design changes in a lab notebook.

Tech Note
The Serial Monitor is a tool used to display electrical data of all types of sensors wired to the Arduino microcontroller.

The block diagram in Figure 23-5 shows the electronic component blocks and the data flow for the Pocket Stage Light. Also, the equivalent Fritzing electronic circuit schematic diagram of the portable lighting device is shown in Figure 23-6. Circuit schematic diagrams allow electronic devices to be built quickly. Electrical/electronic engineers and technicians use them to design, build, and test cool interactive electronic products for games, testing equipment, robots, and automobiles.

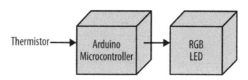

Figure 23-5. *The Pocket Stage Light block diagram*

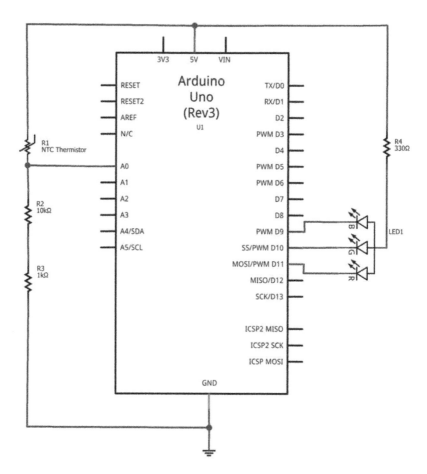

Figure 23-6. *The Pocket Stage Light Fritzing circuit schematic diagram*

Something to Think About

Does a 10KΩ thermistor have a faster RGB LED turn-on response compared to the Ultimate Microcontroller Pack's sensing component?

Electronic Pixel

24

Serial Communications

In nearly all the cool video games you play, the graphics are complex and sophisticated. The images seem real, and their movement is smooth and natural. The magic behind these cool computer graphic images is just a dot of light called a *pixel*, short for "picture element." The game you're playing calculates what the color value of each of the pixels on your screen should be; together, these pixels are arranged to create an image. This project is going to be just like taking one pixel from a monitor and controlling its state using a computer and an Arduino microcontroller. The electronic components to build the pixel project are shown in the Parts List. The assembled Electronic Pixel is shown in Figure 24-1.

Parts List

- Arduino microcontroller
- MakerShield kit
- R1: 330Ω resistor (orange, orange, brown stripes)
- S1: DPDT (double pole, double throw) switch
- LED1: RGB LED

Figure 24-1. *The assembled Electronic Pixel*

Let's Build an Electronic Pixel

In the case of the Electronic Pixel, the LED on and off commands are sent from the Arduino's Serial Monitor and converted into equivalent voltage pulses. These voltage pulses are sent through a USB cable attached between the computer and the Electronic Pixel. Digital pin D9 of the Arduino microcontroller is used to turn on and off the RGB LED.

The Electronic Pixel is built using a breadboard with the components wired to each other, as shown in Figure 24-2. Although the Fritzing wiring diagram shows the Electronic Pixel built on a breadboard, the MakerShield protoboard can be used as well. Also, the Fritzing wiring diagram shows a single pole, double throw (SPDT) switch instead of the double pole, double throw (DPDT) electrical component shown in the Parts List. The remainder of the DPDT switch can be wired as shown in Figure 24-2. Refer to Chapter 5 for additional instructions on how to set up the DPDT switch for breadboarding.

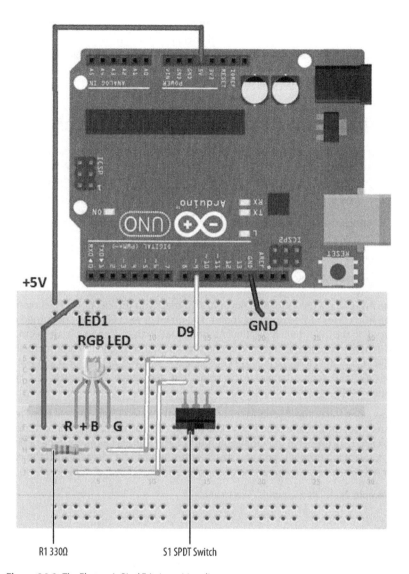

Figure 24-2. *The Electronic Pixel Fritzing wiring diagram*

Tech Note

Remember from an earlier chapter that on and off voltage pulses are known as binary data. The two numbers that represent binary data are 1 and 0. The electrical voltage value for binary 1 is +5 volts and 0 is equal to 0 volts.

Upload the Electronic Pixel Sketch

Before uploading Example 24-1 to the Arduino, check and correct any wiring errors on your breadboard using the Fritzing diagram shown in Figure 24-2. With the Electronic Pixel wired on the breadboard, it is now time to upload the Arduino sketch. Here are the steps you'll need to follow:

1. Attach the Arduino microcontroller to your computer using a USB cable.
2. Open the Arduino software and type Example 24-1 into the software's text editor.
3. Upload the sketch to the Arduino microcontroller.

The Arduino microcontroller is now programmed with the Electronic Pixel sketch. The RGB LED is turned on at this point. The green or red LED might be turned on based on the DPDT switch position. As mentioned in the introduction, an electrical switch is wired to the Arduino microcontroller's digital pin to create a cool interactive effect using the colors of the RGB LED. Slide the switch back and forth and watch the RGB LED toggle between the colors red and green (see Figure 24-3 and Figure 24-4).

Now, open the Arduino Serial Monitor and type the letter "L" into the text box. Press Enter on your computer. The RGB LED is turned off. Type the letter "H" into the text box and click the Serial Monitor "send" button. The RGB LED turns on. If the project is not working properly, find any sketch errors and correct them. Upload the corrected sketch to the Arduino microcontroller and retest the Electronic Pixel.

Example 24-1. *The Electronic Pixel sketch*

```
/*
  Electronic Pixel

  An example of using an Arduino microcontroller for serial communication to
  receive binary data from a computer.  In this case, the Arduino boards
  turns on an RGB LED when it receives the 'H' character, and turns off
  the RGB LED when it receives the 'L' character. Also, an electrical
  switch can change the colors of the RGB LED between green and red.

  The on and off command data can be sent from the Arduino
  Serial Monitor.

*/
```

```
int ledPin = 9;      // the pin that the RGB LED is attached to
int incomingByte;    // a variable to read incoming serial data

void setup() {
  // initialize serial communication:
  Serial.begin(9600);
  // initialize the RGB LED pin as an output:
  pinMode(ledPin, OUTPUT);
}

void loop() {
  // see if there's incoming serial data:
  if (Serial.available() > 0) {
    // read the oldest byte in the serial buffer:
    incomingByte = Serial.read();
    // if it's a capital H, turn on the LED:
    if (incomingByte == 'H') {
      digitalWrite(ledPin, LOW);
    }
    // if it's an L, turn off the LED:
    if (incomingByte == 'L') {
      digitalWrite(ledPin, HIGH);
    }
  }
}
```

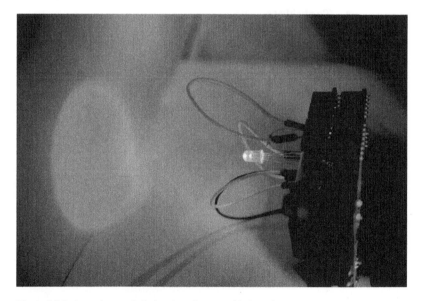

Figure 24-3. *A very large red pixel projected onto a whiteboard*

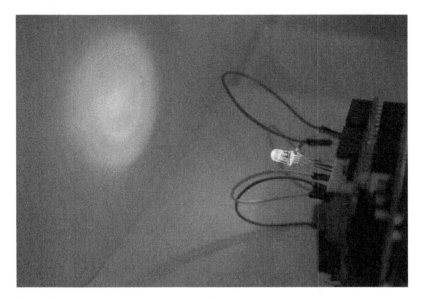

Figure 24-4. *A green ghost projected onto a whiteboard*

 Tech Note

Any computer keyboard characters can be used to operate the RGB LED. Look at Example 24-1 and replace the letters "H" and "L" with the different keyboard characters.

The block diagram in Figure 24-5 shows the circuit component blocks and the data flow for the Electronic Pixel. Also, the equivalent Fritzing electronic circuit schematic diagram of the Electronic Pixel is shown in Figure 24-6.

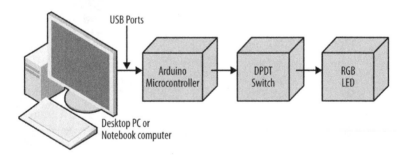

Figure 24-5. *The Electronic Pixel block diagram*

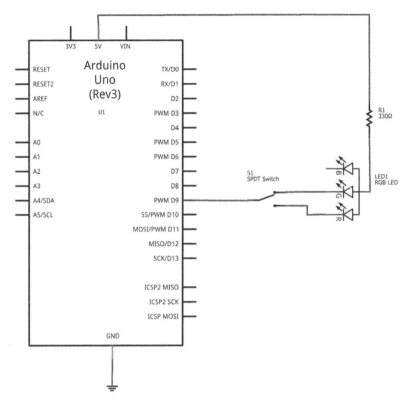

Figure 24-6. *The Electronic Pixel Fritzing circuit schematic diagram*

Something to Think About

How can the switching sequence between the red, green, and blue LEDs be changed to operate faster?

The Metronome

25

The "tick-tick-tick" of the metronome is the steady sound used to help musicians play music at a regular beat. In 1815, Johann Maelzel, a German inventor and engineer, made a swinging device for use as a tool for musicians to keep a steady tempo as they play their music. The metronome is a mechanical gadget that uses an adjustable weight attached to a pendulum rod. As the pendulum rod swings back and forth, the metronome makes a clicking sound. Today, Makers have made a variety of electronic metronomes that use speakers or LEDs to imitate the click and swinging motion of the mechanical pendulum. You can make your own metronome to impress family, friends, and the local Makerspace using a few electronic and electromechanical components from the Ultimate Microcontroller Pack. The Metronome is an awesome project to build because of the cool ticking sound and the homebrew pendulum rod swinging motion. The electronic and electromechanical components to build the Metronome project are shown in the Parts List. The assembled Metronome is shown in Figure 25-1.

Parts List

- Arduino microcontroller
- MakerShield kit
- R1: 10KΩ potentiometer
- P1: piezo buzzer
- M1: DC servo motor

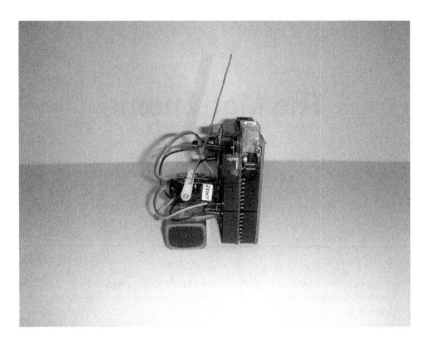
Figure 25-1. *The Metronome*

Let's Build a Metronome

The Metronome is quite easy to build and it looks and sounds awesome when operating. The DC servo motor provides the swinging motion to a homebrew pendulum rod made from a piece of solid wire. The solid wire is threaded through the bottom hole of a servo arm. To secure the wire to the motor while in motion, the end of the wire passing through the bottom hole is wrapped around the servo arm. To complete the mechanical assembly of the servo motor, the homebrew pendulum rod (the solid wire) is stretched out, as shown in Figure 25-1. Next, the servo motor is attached to the breadboard using a piece of solid wire to prevent it from moving when the pendulum rod is swinging back and forth. The wire is wrapped around the servo motor. The free wire ends on each side of the servo motor are inserted into the breadboard (see Figure 25-2).

Figure 25-2. *Servo motor attachment to breadboard: the free wire ends are inserted into the breadboard*

The 10KΩ potentiometer is used as a volume control to adjust the sound level of the piezo buzzer. A cool trick used to make the "tick" sound, along with adjusting the volume, is to place a small piece of tape over the piezo buzzer. Figure 25-3 shows the location of the volume control, and the piezo buzzer with tape placed over it. The Fritzing wiring diagram for building the Metronome is shown in Figure 25-4. As with previous projects presented in this book, the MakerShield protoboard is a great prototyping tool to use in building this cool mini Metronome device. Its breadboarding area allows the piezo buzzer, potentiometer, and servo motor components to be wired to the Arduino microcontroller in a compact package.

Figure 25-3. *View of 10KΩ potentiometer (volume control) and piezo buzzer (with tape)*

 Tech Note
Make sure when placing the tape over the piezo buzzer that you do not completely silence it!

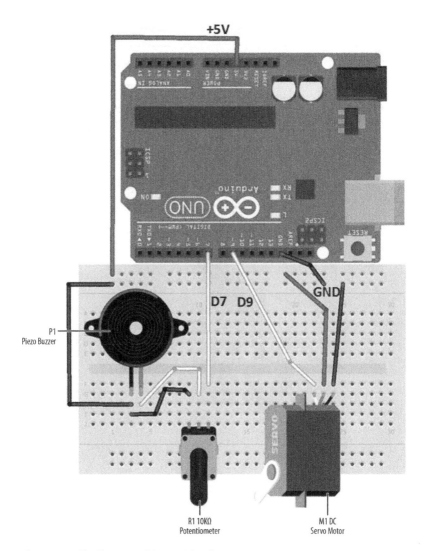

+5V

D7 D9

GND

P1
Piezo Buzzer

R1 10KΩ
Potentiometer

M1 DC
Servo Motor

Figure 25-4. *The Metronome Fritzing wiring diagram*

Upload the Metronome Sketch

Before uploading Example 25-1 to the Arduino, check and correct any wiring errors on your breadboard using the Fritzing diagram shown in Figure 25-4. With the Metronome electrical circuit wired on the breadboard, now it's time to upload the Arduino sketch. Here are the steps you'll need to follow:

1. Attach the Arduino microcontroller to your computer using a USB cable.

2. Open the Arduino software and type Example 25-1 into the software's text editor.

3. Upload the sketch to the Arduino microcontroller.

The Arduino microcontroller is now programmed with the Metronome sketch. The servo motor's arm, with the homebrew pendulum rod, will begin to move back and forth. Also, with each pass of the pendulum rod, a "tick" sound can be heard from the Piezo buzzer. Figure 25-5 shows the completed Metronome in action.

Figure 25-5. *The Metronome in action*

Tech Note

A piece of solid wire was used to make a pendulum rod. Cardboard and LEGO brick plates are good substitutes for a pendulum rod as well.

Example 25-1. *The Metronome sketch*

```
/*
  Metronome sketch

  The servo motor arm will swing back and forth with a tick sound coming
  from a piezo buzzer.

  31 August 2013
  by Don Wilcher

*/




#include <Servo.h>

Servo myservo;   // create servo object to control a servo
                 // a maximum of eight servo objects can be created

int pos = 0;     // variable to store the servo position
int PBuzzer = 7; // piezo buzzer pin number

void setup()
{
  myservo.attach(9);  // attaches the servo on pin 9 to the servo object
  pinMode(PBuzzer, OUTPUT);
}


void loop()
{
  for(pos = 0; pos <=45; pos += 1)  // goes from 0 degrees to 45 degrees
  {                                 // in steps of 1 degree
    if(pos==45){
      digitalWrite(PBuzzer, LOW);
      delay(15);
      digitalWrite(PBuzzer, HIGH);
      delay(15);
      digitalWrite(PBuzzer, LOW);
      delay(15);
    }
    myservo.write(pos);       // go to position in variable 'pos'
    delay(15);                // waits 15ms to reach the position
  }

    for(pos = 45; pos>=1; pos-=1)     // goes from 45 degrees to 0 degrees
    {
    if (pos==1){
      digitalWrite(PBuzzer, LOW);
      delay(15);
      digitalWrite(PBuzzer, HIGH);
      delay(15);
      digitalWrite(PBuzzer, LOW);
```

```
      delay(15);

   }
   myservo.write(pos);    // go to position in variable 'pos'
   delay(15);             // waits 15ms to reach the position
  }
}
```

The Metronome's block diagram is shown in Figure 25-6. The block diagram shows the circuit component blocks and the electrical signal flow (current) for the Metronome. Also, the equivalent Fritzing electronic circuit schematic diagram of the Metronome is shown in Figure 25-7. Electrical/electronic engineers and technicians use them to design, build, and test cool interactive electronic products for games, testing equipment, robots, and automobiles.

Tech Note
In order for the tick sound and swinging motion to be synchronized, the delay(15) instruction is used throughout the Metronome sketch.

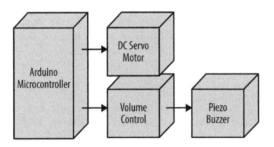

Figure 25-6. *The Metronome block diagram*

Something to Think About

How can you change the timing of the Metronome?

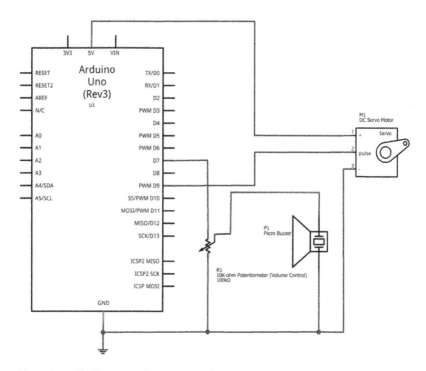

Figure 25-7. *The Metronome Fritzing circuit schematic diagram*

The Secret Word Game

In the mid-1970s, a company named Tiger Electronics made a variety of handheld electronic sporting games like baseball, basketball, and football. These portable handheld games used LEDs, seven-segment LED displays, and speakers to provide visual and audio effects for the units. The pushbuttons on these portable gaming devices allowed total interaction between the player and the electronic handheld unit. In other words, while they weren't a Wii, these little electronic units were the first really popular interactive sports games.

Imagine being able to make your own retro portable electronic game using components from the Ultimate Microcontroller Pack. The Secret Word game is a cross between *Jeopardy!* and Charades, in which players attempt to guess the message programmed into the Arduino microcontroller within a certain amount of time. The rules of the game are quite simple and will be discussed later in this final chapter of the book. The assembled Secret Word Game is shown in Figure 26-1.

Parts List

- Arduino microcontroller
- Full-size breadboard
- R1: 330Ω resistor (orange, orange, brown stripes)
- R2: photocell (light sensor)
- R3: 1KΩ resistor (brown, black, red stripes)
- LCD1: 16x2 liquid crystal display (LCD), part number JHD 162A
- LED1: RGB (red, green, blue) LED
- S1: pushbutton switch

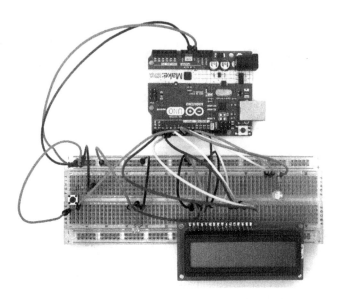

Figure 26-1. *The Secret Word Game*

Let's Build a Secret Word Game

The Secret Word Game is a little tricky to build because of the wiring. Therefore, you'll have to use the full breadboard that comes with the Ultimate Microcontroller Pack to adequately space the parts, as shown in Figure 26-1. Use the Fritzing wiring diagram shown in Figure 26-2 to build the Secret Word Game on the full-size breadboard.

Pin 1 of the LCD is the leftmost input at the base of the screen. Pins 2 to 16 continue to the right. (Another way to identify pin 1 is by the small circle placed on the PCB right next to pin 1.)

The photocell and RGB LED should be placed on the breadboard so that they are easily visible and accessible; you need to clearly see the LED, and easily shine a light on the photocell.

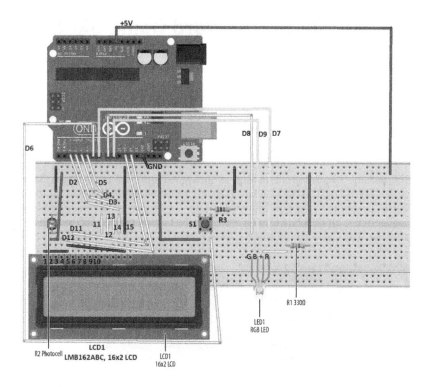

S1 = Start Game Pushbutton
Switch

R3 = 1KΩ

Figure 26-2. *The Secret Word Game Fritzing wiring diagram*

In past projects, we used the smaller sized MakerShield to keep the projects compact. But with this many electronic components and wires in the Secret Word Game, the full-size breadboard is the best choice for this project. And as always, be sure to recheck your wiring against the Fritzing diagram, to catch any possible errors.

Tech Note

Before you upload the Arduino sketch to the Secret Word Game, have a set of fresh eyes (like a friend or parent) look at the electrical wiring on the full-size breadboard to catch any mistakes you might have missed.

Upload the Secret Word Game Sketch

With the Secret Word Game wiring on the breadboard completed, now it's time to upload the Arduino sketch. Here are the steps you'll need to follow:

1. Attach the Arduino microcontroller to your computer using a USB cable.

2. Open the Arduino software and type Example 26-1 into the software's text editor.

3. Upload the sketch to the Arduino microcontroller.

The Arduino microcontroller is now programmed with the Secret Word Game sketch. The LCD will be blank and the RGB LED turned off. When you press the "Start Game" pushbutton, the RGB LED will light up. The red, green, and blue LEDs will sequence five times before turning off. Figure 26-3 shows the RGB LED sequencing after the Start Game pushbutton has been pressed.

Figure 26-3. *The Secret Word Game starting its timing sequence using the RGB LED*

Once the RGB LED has turned off, shining a light on the photocell will reveal the secret word on the LCD (Figure 26-4). Removing the light from the photocell will erase the secret word on the LCD. New secret words can easily be uploaded to the Arduino by changing one line of instruction in the sketch.

Figure 26-4. *The secret word "Cat" being revealed on the LCD*

Example 26-1. *The Secret Word Game sketch*

```
/*

    Demonstrates the use of a 16x2 LCD. A brief press of the Start Game
    pushbutton will turn on the RGB LED timing sequencing. The RGB LED turns
    off and the secret word can be revealed by a shining light on a photocell.

    25 August 2013
    by Don Wilcher

    */

// include the library code:
#include <LiquidCrystal.h>

// initialize the library with the numbers of the interface pins
LiquidCrystal lcd(12, 11, 5, 4, 3, 2);
int buttonPin = 6;      // the number of the Start Game pushbutton pin
int RPin = 7;           // select the pin for the red LED
int GPin = 8;           // select the pin for the green LED
int BPin = 9;           // select the pin for the blue LED

// variables will change:
int buttonStatus = 0;   // variable for reading the Start Game
                        // pushbutton status

void setup() {
  // initialize the pushbutton pin as an input:
  pinMode(buttonPin, INPUT);

  // declare the LED pins as outputs:
  pinMode(RPin, OUTPUT);
  pinMode(GPin, OUTPUT);
  pinMode(BPin, OUTPUT);
```

```
  // set up the LCD's number of columns and rows:
  lcd.begin(16, 2);

}

void loop() {
  // read the state of the pushbutton value:
  buttonStatus = digitalRead(buttonPin);
  // check if the pushbutton is pressed
  // if it is, the buttonState is HIGH:
  if (buttonStatus == HIGH) {
    lcd.clear();
    delay(500);
    for (int i=0; i <= 5; i++){
      lcd.setCursor(8,0);
      lcd.print(i);

      // turn the red LED on:
      digitalWrite(BPin, HIGH);
      digitalWrite(RPin, LOW);

      // delay red LED for 1/2 second:
      delay(500);
      // turn the green LED on:
      digitalWrite(RPin, HIGH);
      digitalWrite(GPin, LOW);
      // delay green LED for 1/2 second:
      delay(500);
      // turn the blue LED on:
      digitalWrite(GPin, HIGH);
      digitalWrite(BPin, LOW);
      //delay blue LED for 1/2 second:
      delay(500);
    }
  } else {
      //turn red, green, and blue LEDs off:
      digitalWrite(RPin, HIGH);
      digitalWrite(GPin, HIGH);
      digitalWrite(BPin, HIGH);

      // print a Secret Word to the LCD:
      lcd.setCursor(0,0);
      lcd.print("Secret Word is:");
      // set the cursor to column 0, line 1
      // (note: line 1 is the second row, since counting begins with 0):
      lcd.setCursor(0, 1);
      // print the number of seconds since reset:
      lcd.print("Cat"); // change secret word or phrase here!
  }
}
```

Tech Note
An LCD can also print mini messages in addition to words.

The Secret Word Game's block diagram is shown in Figure 26-5. The Fritzing circuit diagram is shown in Figure 26-6. Circuit schematic diagrams allow you to build electronic devices quickly. Electrical/electronic engineers and technicians use them to design, build, and test cool interactive electronic products for games, testing equipment, robots, and automobiles.

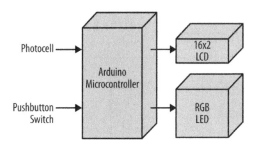

Figure 26-5. *The Secret Word Game block diagram*

Rules for the Secret Word Game

The objective of the game is to guess the mystery word programmed into the Arduino microcontroller within a certain amount of time. Three hints will be given to the players by the game host, after which a pushbutton switch is pressed to start the RGB LED timing sequence. The players will write their answers on a sheet of paper. Once the RGB LED timing sequence is completed, the game host reveals the secret word to the players by shining a flashlight on the photocell. The player who has the winning word will shout out, "I'm a Winner," or some other suitably cool phrase! The game host starts a new game by changing the secret word using the lcd.print() instruction of the Secret Word Game sketch. The new sketch is then uploaded to the Arduino microcontroller for the next round of electronic gaming fun!

Something to Think About

How can a mystery phrase be programmed for the game?

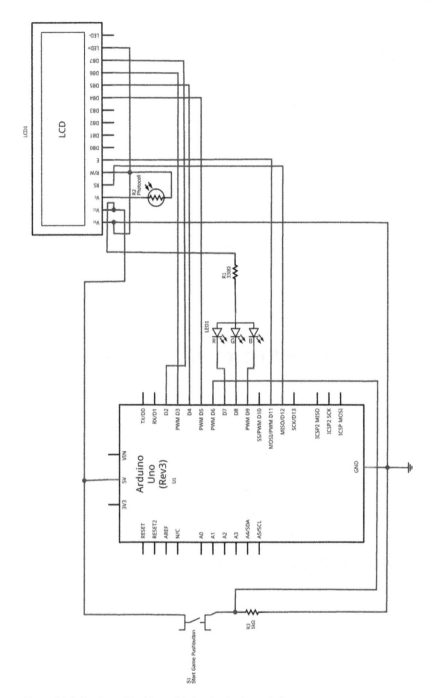

Figure 26-6. *The Secret Word Game Fritzing circuit schematic diagram*

Index

R

resistance
  measuring for electronic components, 111
  mesuring with Arduino Ohmmeter, 113
  reading for tilt control switch, 26
  relationship with voltage, 115
resistor-capacitor timing basics
  Sunrise-Sunset Light Switch, 9–16
  Trick Switch, 1–6
resistors, 111
  connected in series, resistance reading from, 115
  in logic tester with an LCD, 139
retro portable electronic games, 231
  (see also Secret Word Game)
RGB Flasher, 79
  (see also Multicolor RGB Flasher)
  block diagram, 79
  building, 81
  sketch for, 83
RGB LEDs, 79
  common anode, 81
  creating white light with, 208
  in Electronic Pixel, 214
  in Pocket Stage Light, 206
  logic tester with, 131–136
    pinout of RGB LED, 132
  operation with mini pushbutton switch in Magic Light Bulb, 89
  SPST switches controlling, 80
  taking pin LOW to light it, 85
  typical, with pinout names, 80
Rocket Launching Game (with Processing), 169–181
  block diagram and circuit schematic diagram, 181
  building a Rocket Game, 170
  DisplayItems Processing Sketch, 179
  MultiDigital4 sketch, 172–174
  Rocket Game Processing sketch, 174–179
  Rocket Launcher parts list, 169
  uploading MultiDigital4 sketch, 171

S

Secret Word Game, 231–237
  block diagram and circuit schematic diagram, 237
  building, 232
  parts list, 231
  rules for, 237
  uploading the sketch, 234–237
sensors, 19

serial communications (see Electronic Pixel)
Serial Monitor, 210
  using to debug code, 108
serial monitors
  output for tilt control switch information, 24
  Sunrise-Sunset detector with, 14
    displaying Sunset and Sunrise messages, 15
series circuit, 53
servo motors, 197
  build process for Tilt Sensing Servo Motor Controller, 20
  in Metronome, 222
  testing limits of, 193
  tilt sensing servo motor controller, 19
sketches (code)
  Adjustable Twin LED Flasher, 35, 39
  Amazing Pushbutton sketch, 146
  Arduino AND Logic Gate, 57
  Arduino NOT Logic Gate, 46
  Arduino Ohmmeter sketch, 113
  Arduino OR logic gate, 66
  Blink sketch for Twin LED Flasher, 31
  DisplayItems Processing sketch, 152–153, 165, 179
  Electronic Cricket sketch, 202
  Electronic Pixel, 216
  LCD News Reader, 122–128
  Logic Tester (with an LCD), 140
  Logic Tester sketch, 133
  Magic Light Bulb running through tricolor pattern, 89
  Metal Checker sketch, 98
  Metronome sketch, 226
  MultiDigital4 sketch, 172–174
  NTC Sensor Processing sketch, 188
  pa_Pushbutton Processing sketch, 149–152
  pa_Tilt Processing sketch, 162–165
  Pocket Stage Light sketch, 208
  Processing sketch listings for Arduino projects, 162
  Pushbutton, 3
  Pushbutton with LED indicators changes, 5
  RGB Flasher, 83
  Rocket Game Processing sketch, 176–179
  Secret Word Game, 234–237
  Sunrise Sunset Detector with Serial Monitor, 14
  Sunrise-Sunset Light Switch, 11
  Sweeping sketch, 196
  Temperature Indicator sketch, 186

About the Author

Don Wilcher is a passionate teacher of electronics technology and an electrical engineer with 26 years of experience. He's worked on industrial robotic systems, automotive electronic modules and systems, and embedded wireless controls for small consumer appliances. While at Chrysler Corporation, Don developed a weekend enrichment pre-engineering program for inner-city kids. He is an Electronics and Robotics Technologist developing twenty-first century educational products for Makers and educators.

The cover and body font is Benton Sans, the heading font is Serifa, and the code font is Bitstreams Vera Sans Mono.